T0253565

Computer Modeling and Simulation of Dynamic Systems Using Wolfram SystemModeler

Kirill Rozhdestvensky · Vladimir Ryzhov ·
Tatiana Fedorova · Kirill Safronov ·
Nikita Tryaskin · Shaharin Anwar Sulaiman ·
Mark Ovinis · Suhaimi Hassan

Computer Modeling and Simulation of Dynamic Systems Using Wolfram SystemModeler

 Springer

Kirill Rozhdestvensky
Department of Applied Mathematics
and Mathematical Modeling
St. Petersburg State Marine Technical
University (SMTU)
Saint Petersburg, Russia

Tatiana Fedorova
Department of Applied Mathematics
and Mathematical Modeling
St. Petersburg State Marine Technical
University (SMTU)
Saint Petersburg, Russia

Nikita Tryaskin
Department of Hydromechanics
and Marine Acoustics
St. Petersburg State Marine Technical
University (SMTU)
Saint Petersburg, Russia

Mark Ovinis
Department of Mechanical Engineering
Universiti Teknologi PETRONAS
Seri Iskandar, Malaysia

Vladimir Ryzhov
Department of Applied Mathematics
and Mathematical Modeling
St. Petersburg State Marine Technical
University (SMTU)
Saint Petersburg, Russia

Kirill Safronov
Department of Applied Mathematics
and Mathematical Modeling
St. Petersburg State Marine Technical
University (SMTU)
Saint Petersburg, Russia

Shaharin Anwar Sulaiman
Department of Mechanical Engineering
Universiti Teknologi PETRONAS
Seri Iskandar, Malaysia

Suhaimi Hassan
Department of Mechanical Engineering
Universiti Teknologi PETRONAS
Seri Iskandar, Malaysia

ISBN 978-981-15-2802-6   ISBN 978-981-15-2803-3   (eBook)
https://doi.org/10.1007/978-981-15-2803-3

This Springer imprint is published by the registered company Springer Nature Singapore Pte Ltd.
The registered company address is: 152 Beach Road, #21-01/04 Gateway East, Singapore 189721, Singapore

The textbook was prepared with the financial support of grant No. 573751-EPP-1-2016-1-DE-EPPKA 2-CBHE-JP, InMotion: New strategies for training engineers using visual modeling environments and open training platforms implemented under the program Erasmus+, capacity building in higher education.

The support of the European Commission in writing this work does not include agreement with the content, which reflects only the point of view of the authors. The commission is not responsible for any use of this information.

Co-funded by the
Erasmus+ Programme
of the European Union

InMotion

Reviewers

Dr. Sharul Sham Dol, Department of Mechanical Engineering, College of Engineering, Abu Dhabi University, PO Box 59911 Abu Dhabi, UAE;

Dr. Muhammad Yasin Naz, Department of Physics, University of Agriculture, 38040 Faisalabad, Pakistan.

Kirill Rozhdestvensky, Vladimir Ryzhov, Tatiana Fedorova, Kirill Safronov, Nikita Tryaskin, Shaharin Anwar Sulaiman, Mark Ovinis, Suhaimi Hassan

Computer Modeling and Simulation of Dynamic Systems Using Wolfram SystemModeler: Textbook by Kirill Rozhdestvensky, Vladimir Ryzhov, Tatiana Fedorova, Kirill Safronov, Nikita Tryaskin, Shaharin Anwar Sulaiman, Mark Ovinis, Suhaimi Hassan.

The textbook «Computer modeling and simulation of dynamic systems using Wolfram SystemModeler» briefly describes the basic concepts of the theory of modeling and the methods of constructing computer models of dynamic systems using an interactive graphical modeling and simulation environment Wolfram SystemModeler, as well as illustrative examples of solving problems in mechanics and hydraulics.

The first chapter introduces the reader to the basic concepts of mathematical modeling. It discusses the important concepts such as a model, physical, abstract modeling and simulation and gives an idea of the classification and basic properties of mathematical models. The computer simulation process is divided into stages, and a brief description of each stage is given. Particular attention is paid to identifying differences between continuous, discrete and hybrid models. The second chapter describes Wolfram SystemModeler (WSM). The WSM is intended for computer simulation of complex multidomain physical and engineering systems and processes based on the Modelica language. It allows to simulate the developed models and provides a wide range of tools for analyzing the results of the simulation. The reader gets acquainted with the basics of writing program code with use of the object-oriented language, Modelica. The basics of working in the ModelCenter to create a component model using the Modelica standard library are demonstrated. The rules for performing a computational experiment in the

Simulation Center are described in detail. The third chapter gives a general idea of the fundamental principles that underlie any mathematical model such as conservation laws or variational principles. On these principles, the simplest models are built. These models are used as the basis for building more complex hierarchical models. Application of fundamental principles is illustrated with examples such as a harmonic oscillator. The reader gets acquainted with the hierarchical principle of building models, from simple to complex. In the fourth chapter, the reader is invited to study methods of modeling single component, multicomponent and hybrid systems using two alternative approaches: developing computer models directly with Modelica language and using component modeling technologies with application of the standard Modelica libraries. The following mechanical oscillatory systems with one degree of freedom are considered: a mathematical pendulum, a Galileo's pendulum, an elastic pendulum. In the fifth chapter, the methods for solving more complex problems are discussed. These are multicomponent and hybrid systems. It is proposed to use two approaches: directly developing computer models on the Modelica language and using component modeling technologies with application of the standard Modelica libraries. The modeling of mechanical oscillatory systems with several degrees of freedom is considered. The following illustrative examples of such dynamic models are given: the relative motion of two bodies, complex oscillatory mechanical systems of dampers and springs, and coupled and double mathematical pendulums. The sixth chapter is devoted to building hierarchical component models and creating custom libraries in the Wolfram SystemModeler. The reader gets acquainted with the concepts of blocks and connections. The concepts of oriented and non-oriented connections between blocks are introduced. Mathematically, these connections include algebraic and differential equations which describe the behaviors and interactions of system objects. The following illustrative examples of such models are demonstrated: model of draining of a tank; problem of heating a liquid mixture with the help of a PID controller (by using the custom components); inverted pendulum problem.

Introduction

At present, to study complex processes and systems, along with a full-scale experiment, a computational experiment is used—a methodology based on the application of applied mathematics and computer technology, combining the advantages of both theory and experiment. A computational experiment has a number of indisputable advantages over a full-scale one. The main specificity of the computational experiment (computer simulation) is the ability to virtually predict the properties of the object being modeled, depending on the various conditions of its operation and variable design parameters. In addition, for the study of complex objects and systems, computational experiment is less expensive, and its implementation requires significantly less time. In addition, it can predict the results of critical tests, which may, in reality, lead to the loss of the full-scale object under investigation. A computational experiment acquires exceptional importance in those cases when full-scale experiments or physical model experiments are impossible. The adequacy of computer simulation results is determined by the credibility of the employed mathematical model which describes the system or process with a given accuracy.

The construction of a mathematical model that allows to obtain correct results in the process of computer simulation includes the following steps:

- building a model that involves the separation of all factors acting in the phenomenon under consideration into major and minor factors, the latter is not accounted for at the initial stage of research; formulation of assumptions and conditions of applicability of the model; determination of the boundaries in which the results would be fair; describing the model in mathematical terms, usually, in the form of differential or integro-differential equations;
- development (selection) of a method for solving a posed mathematical problem. When building a method for solving a problem, the accuracy and efficiency of the numerical methods used for calculation are important;
- development of an algorithm for solving the problem. When implementing this stage, it should be remembered that the accuracy of calculations on a computer is limited; in addition, any method of calculation also introduces an error.

Selected or created computational algorithms, as well as the introduction of simplifying assumptions into the model, should not substantially distort the basic properties of the object under study, and they should be adaptable to the peculiarities of the tasks being solved and the computational tools used;

– creation of software for the implementation of the model and algorithm on the computer. Software must take into account the specifics of mathematical modeling associated with the use of a hierarchy of mathematical models and multivariate calculations. This fact implies the possibility of using software packages developed, in particular, using object-oriented programming;

– carrying out a computational experiment, which allows to obtain systematic calculated results determined by various input parameters of the problem;

– processing of calculation results, their analysis, comparison (if possible) with the results of a full-scale experiment, formulation of conclusions, and recommendations. It should be noted that at this stage it may be necessary to clarify the mathematical model in order to obtain an adequate practical result.

The methodological universality of computer modeling makes it possible, on the basis of accumulated experience in the development of mathematical models, numerical methods, computational algorithms, and means of analyzing the results obtained, to quickly and effectively solve various applied problems.

In practice, for the study of complex multicomponent systems, a computational experiment is often carried out using specialized software packages. When solving problems of a specific domain, certain requirements arise for the packages used.

The textbook discusses the modeling of dynamic systems (class of tasks related to the macro-level). We study problems whose mathematical models are described by differential and differential–algebraic equations. To solve this class of problems, computer math packages such as MATLAB, Simulink [1], MapleSim [2], Rand Model Designer [3], ISMA [4], and Wolfram SystemModeler [5] can be used.

Each of these packages, having powerful functionality, including visual modeling tools, can best solve specialized applied tasks of a certain type. In this regard, when solving a specific engineering problem, the question arises which of the packages of computer mathematics would be more effective for studying a particular problem. To answer this question, you must either have experience with each of the packages or have information about the features of these packages. There are works devoted to the comparative analysis of visual modeling tools, thanks to which the user can decide on the choice of a tool according to his needs.

In the textbook, the Wolfram SystemModeler software environment [5] is considered as a tool for computer modeling. A review of the functionality of the package is made, its specificity is discussed in comparison with other packages of computer mathematics, and the possibilities of the package for the solution of applied engineering problems of mechanics are considered.

Wolfram SystemModeler is an interactive graphical environment designed for mathematical and computer simulation of multi-dimensional systems using the Modelica language.

Wolfram SystemModeler allows the user to create models of systems, both independently and with the help of an extensive library of physical and logical components. The ability to integrate with Mathematica [6] allows to improve the visualization of processes when solving complex engineering problems, as well as to support virtual modeling with additional mathematical calculations.

Speaking about the functionality of SystemModeler, the following can be noted:

- drag-and-drop approach for complex systems modeling;
- use of Modelica object-oriented language as a basis for modeling;
- management of computational experiments in the Simulation Center interactive environment using the Wolfram Language;
- easily implemented animation;
- text user interface for dynamic system modeling and their analysis based on differential equations;
- acausal, i.e., component-based and casual, block-based modeling;
- 2D and 3D modeling of mechanical systems, electrical processes, hydraulic systems, thermodynamic processes, control systems, etc.;
- frequency analysis, sensitivity analysis with respect to input parameters, and analysis of the reliability of the created system;
- integration with Mathematica for analyzing and storing the created models.

When creating models in the Wolfram SystemModeler environment, Modelica is used—a freely distributed object-oriented, declarative, multi-domain modeling language for component-oriented modeling of complex systems, in particular, systems containing mechanical, electrical, electronic, hydraulic, thermal, energy components, and also control components.

Modelica language is based on writing differential, algebraic, and discrete equations, instead of using assignment operations. Such a modeling method does not specify a predetermined causal relationship to the calculation of input variables. The Modelica language compiler independently manipulates the equations in symbolic form, determining the order of their execution and which components in the equation will determine the inputs and outputs. Thus, the program in Modelica is a system of equations.

Modelica is often compared with object-oriented programming languages, such as C++ or Java, but in fact it differs significantly from them. The first and most obvious difference is that Modelica is a modeling language, not a programming language. Classes of this language are not compiled in the usual sense, but are converted into objects, which are subsequently used by a specialized process. The second difference, which was already mentioned in the text, is that classes can contain algorithmic components that are similar to operators and blocks in programming languages. The Modelica language declares that a class is any definition, including an algorithmic function. This language is suitable and is used for software and hardware modeling and for modeling embedded control systems.

The textbook consists of six chapters and has the following structure.

The first chapter introduces the reader to the basic concepts of mathematical modeling. It introduces important concepts such as a model and types of modeling and gives an idea of the classification and basic properties of mathematical models. Special attention is paid to identifying differences between continuous, discrete, and hybrid models.

The second chapter is devoted to the description of the Wolfram SystemModeler environment. The basics of working in the Model Center for creating your own model with components from the Modelica standard library, principles for creating your own components, and rules for performing a computational experiment in the Simulation Center are given.

The third chapter gives a general idea of the fundamental principles underlying any mathematical model, such as the laws of conservation in classical mechanics or variational principles. On these principles, the so-called basic models can be built, which later serve as the foundation on which more complex hierarchical models are built.

In the fourth, fifth, and sixth chapters, the reader is invited to explore methods for solving single component, multicomponent, and hybrid problems using two alternative approaches: direct programming of mathematical models in Modelica [7] and component modeling technologies using standard libraries [5]. In the last three chapters, illustrative examples of constructing models of dynamic systems are given with increasing degree of complexity.

The knowledge acquired through this textbook can be used by students for independent assignments and virtual laboratory work.

In preparing the tutorial, materials from the site of the software development company Wolfram Research Inc. [8] were used.

References

1. Matlab, Simuink: https://www.mathworks.com/
2. Maple: https://www.maplesoft.com/
3. Rand Model Designer: https://www.mvstudium.com/
4. ISMA: https://drive.google.com/open?id=1xV25DkoYK_Qnp4oKIOob2lNbn3rVM9K5
5. Wolfram SystemModeler: https://www.wolfram.com/system-modeler/
6. Mathematica: https://www.wolfram.com/mathematica/
7. Modelica: https://www.modelica.org/
8. Wolfram Research Inc.: https://www.wolfram.com/

Contents

Chapter 1
Modeling Systems

1.1 Basic Modeling Concepts

The decisive role in scientific research is played by a full-scale experiment and observation of physical objects. However, in certain cases, a full-scale experiment may be difficult or economically impractical. An alternative to a full-scale experiment can be a model experiment or a modeling process. The purpose of the simulation is to predict with practical accuracy the properties and behavior of the studied real objects by replacing them with simpler model analogs. Thus, modeling, as a more accessible and at the same time effective tool, allows using the constructed model to investigate the characteristics of a complex real object and in some cases even to detect previously unknown effects.

A model can be a physical or abstract analog, which can reproduce the properties of a real object of research with a certain degree of accuracy. At the same time, it should be noted that the adequacy of the results of the modeling process to real results can be finally confirmed by a full-scale experiment.

Simulation can be used not only for research purposes related to predicting the characteristics of an object depending on the given conditions and modes of its operation (which, unlike the case of a full-scale experiment, can be easily changed), but also for learning to work with any complex systems or objects.

In the latter case, the model plays the role of simulating the functioning of a real object. An example is simulation systems (simulators) that allow controlling the motion of a vessel using a virtual dashboard. It is obvious that the use of a simulator for training personnel to work in real conditions is advisable, since it allows one to simulate critical situations, the appearance of which during the operation of a real vessel is unacceptable.

The above suggests that modeling technologies have certain advantages compared with direct experimental studies of real objects or processes. They can be effective tools for exploring the world. However, it should be understood that the modeling

© Springer Nature Singapore Pte Ltd. 2020
K. Rozhdestvensky et al., *Computer Modeling and Simulation of Dynamic Systems Using Wolfram SystemModeler*,
https://doi.org/10.1007/978-981-15-2803-3_1

process can be very complex and requires in-depth knowledge in a specific subject area.

Modeling can be divided into physical and abstract.

Physical modeling suggests that some other system with the same physical nature is used as the source object model. This system is created on the basis of the similarity theory, which allows us to state that the model retains the required properties of the original object. In engineering practice, a prototype of an object is produced, and tests are conducted, during which its output parameters and characteristics are determined. The results of such tests can be extended to a real object also considering similarity criteria.

A variety of physical modeling is analog modeling, based on replacement of the original object with an object of a different physical nature, but with similar behavior.

Unlike physical modeling, abstract modeling describes objects based on its abstract images—most often they are various signs, symbols, schemes, graphics, etc.

The most important type of abstract modeling is mathematical modeling based on the use of mathematical tools and mathematical logic. Mathematical modeling is the process of establishing the correspondence between the original real system and a certain mathematical model, as well as the study of this model, which allows one to evaluate the characteristics of the real system. To analyze the results of mathematical modeling, an interpreter is required, which may be a professional in a specific subject area or a computer.

For the mathematical model, it is a characteristic that the processes of functioning of the system under study are described in the form of some functional relationships (algebraic, differential, integral, and other equations) and logical conditions. If the set of parameters characterizing the model can be explicitly expressed from functional dependencies by analytical methods, then it is said that an analytical solution can be obtained. In the case when this is not possible, a solution can only be obtained by using numerical methods.

The analytical solution can be obtained only for relatively simple systems. For complex systems, there are often big math problems. Often, to use the analytical method, the initial model is significantly simplified, which can affect the quality of the solution.

When it is not possible to find a solution to equations in a general form, one can apply a qualitative method, when in the absence of a solution in an explicit form, one can find some of its properties (e.g., estimate the stability of a solution). If, however, it is necessary to obtain a solution, the equations can be investigated by numerical methods for specific initial data. In this case, computer modeling finds great use.

For computer simulation, it is typical to represent the mathematical model of the system in the form of a computational algorithm.

Computer modeling can be divided into numerical, imitation, and statistical.

In numerical simulation, computational mathematics methods are used to build a computer model, and a computational experiment consists of numerically solving mathematical equations for given parameters and initial conditions.

Imitational modeling is a type of computer simulation, which is characterized by reproduction on a computer (imitation) of the process of functioning of the system under study. This simulates the elementary phenomena that make up the process, while preserving their logical structure and the sequence of flow in time, which allows to obtain information about the state of the system at specified points in time.

Statistical modeling is a type of computer simulation that allows to obtain statistical data on the processes in the simulated system.

The stages of computer simulation are: model development, algorithm development, and software implementation.

At the first stage, an "equivalent" object is built. This "equivalent" reflects in a mathematical form, the properties of an object that are important for a given study: the laws to which the object obeys, the connections inherent in its parts, etc. This stage is characterized by the principle of decomposition, that is, the division of the original object into separate elements and the mathematical description of each element. Then, the mathematical model can be investigated by theoretical methods, which allows one to obtain preliminary knowledge about the object.

The second stage is the development of an algorithm for implementing the model on a computer. The model is presented in a form convenient for the application of numerical methods, and the sequence of computational and logical operations that need to be performed is determined in order to find the desired quantities with a given accuracy. Computational algorithms should not distort the basic properties of the model and, consequently, the original object, should be economical, and should adapt to the peculiarities of the tasks and computers used.

At the third stage, programs are created that "translate" the model and algorithm into a language accessible to a computer. They are also subject to the requirements of efficiency and adaptability. They can be called the "electronic" equivalent of the object being studied, already suitable for testing on a computer.

The method of computer simulation combines the advantages of theory and experiment. Indeed, working not with the object itself (a phenomenon or process), but with a computer model, makes it possible to investigate its properties relatively quickly, without significant expenses. This constitutes the advantages of the theory. At the same time, computational experiments with models of objects make it possible, based on the power of computational methods and computers, to study objects in great breadth and depth, which is inaccessible by purely theoretical approaches. This already constitutes the advantages of the experiment.

1.2 Classification of Mathematical Models

Mathematical models are distinguished by the nature of the displayed properties of the system, their degree of detail, methods of derivation, and formal presentation.

By the way of presenting the system, the models are divided into structural and functional. If the mathematical model reflects the structure of the simulated system, that is, it highlights the elements and their system connections, then it is called the structural mathematical model. If the mathematical model reflects only how the system works, given its internal structure, then it is called functional. There may also be combined mathematical models that describe both the functional and structural properties of the system.

Functional models can be classified by different properties. The following types of models can be distinguished.

By the set of parameter values, the models are divided into continuous and discrete. Each system parameter can be of two types—continuously changing in a certain interval of its values or accepting only some discrete values. An intermediate situation is also possible, when in one region the parameter takes all possible values and in the other only discrete ones. The same definition can be applied to the description of the behavior of models. If all parameters of the model are continuous, then the model is called continuous; if the parameters are discrete, then the model is discrete. If some parameters can change both continuously and discretely, then this model is hybrid.

By the type of dependencies between the parameters, the models are divided into linear and nonlinear. If all output parameters of the model are linearly dependent on the input (superposition principle), then the model is called linear. In the opposite case, it is referred as a nonlinear model. Linear models are simpler and allow you to get an analytical solution, but they do not always allow satisfactory description of the object under study.

By the relation to external factors, models can be divided into open and closed (isolated). A closed model is a model that functions without communication with external variables. In a closed model, changes in model parameter values in time are determined only by the internal interaction of the parameters themselves. An open model associated with external variables.

By the presence or absence of random parameters in the system, the models are divided into deterministic and stochastic. The deterministic model does not take into account the influence of random factors; therefore, its behavior is predetermined by a system of equations. With repeated experiments with the same initial values, the result will be the same. There are random parameters in the stochastic model, so its behavior cannot be predicted.

The essential feature of the classification of mathematical models is their ability to describe the change in system parameters over time. If the state of the model does not depend on time, then the model is called static. In contrast, a model in which parameters change over time is called dynamic. Separately, it is worth highlighting the class of stationary models. Stationary models describe systems in which so-called steady-state processes occur, i.e., processes in which the output parameters are constant in time. Periodic processes are also established, in which some output parameters remain unchanged and the rest oscillate.

By the degree of detail of the description of the processes occurring in the system, functional mathematical models can be divided into hierarchical levels: micro-, macro-, and meta-level. Mathematical models of the micro-level describe processes in systems with distributed parameters and mathematical models of the macro-level in systems with concentrated parameters. In micro-level models, phase variables can depend on both time and spatial coordinates, whereas in macro-level models, they depend only on time. If in a mathematical model of a macro-level the number of phase variables is of the order of 10^4–10^5, then such mathematical models are referred to as a meta-level. The quantitative analysis of such a mathematical model is very complicated, and it requires a significant amount of computational resources. In this case, by combining and enlarging the elements of a complex system, they strive to reduce the number of phase variables by excluding the internal parameters of the elements from consideration, limiting themselves only to the description of mutual relations between the enlarged elements.

The most common form of representation of a dynamic (evolutionary) mathematical model of a micro-level is the formulation of a boundary value problem for differential equations of mathematical physics. This formulation includes partial differential equations with initial and boundary conditions.

The most common form of representation of a dynamic macro-level model is a model described by a system of algebraic–differential equations.

1.3 Basic Properties of Mathematical Models

The use of mathematical models to study the characteristics of the original object will be effective if the properties of the mathematical model satisfy several requirements [1]. These properties include: completeness, adequacy, accuracy, robustness, productivity, and efficiency.

The completeness of the mathematical model makes it possible to sufficiently reflect those characteristics and features of the system that interest us from the point of view of the goal of the computational experiment. That is, completeness shows how the properties of the model correspond to the properties of the object of study.

The adequacy of a mathematical model is the ability of a mathematical model to reflect the properties of a system with a relative error not worse than a given one. In a general sense, the adequacy of a mathematical model is understood as the correct qualitative and fairly accurate quantitative description of precisely those characteristics of the system that are important in this particular case. A model that is adequate when selecting some characteristics may be inadequate when choosing other characteristics of the system.

The accuracy of the mathematical model makes it possible to provide an acceptable agreement of the real values with the output parameters of the system found using a mathematical model.

The robustness of a mathematical model characterizes its stability with respect to the errors of the original data, the ability to level these errors and prevent them from excessively affecting the result of a computational experiment.

The productivity of the mathematical model is associated with the reliability of the source data. If they are the result of measurements, then the accuracy of their measurements should be higher than for those parameters that are obtained using a mathematical model. Otherwise, the mathematical model will be unproductive and its use for the analysis of a particular system will lose its meaning.

The cost-effectiveness of a mathematical model is estimated by the cost of computing resources (machine time and memory) necessary to implement a mathematical model on a computer. These costs depend on the number of arithmetic operations when using the model, on the dimension of the space of phase variables and other factors.

1.4 Computer Simulation and Computational Experiment

Most mathematical models of practical interest are very complex and do not allow to obtain a solution in analytical form. Solutions for such models are built using mathematical packages that allow performing symbolic calculations. A model built in this way is called a computer. Computer modeling consists in the development of an algorithm (program) describing a model of the system under study. As noted earlier, computer modeling allows you to repeatedly reduce the cost of developing models compared to non-computer modeling methods and conducting full-scale tests. The creation of a computer model is usually done on the basis of a mathematical one. Thus, computer modeling can be considered as a kind of mathematical modeling.

The computer model must have all the properties of the mathematical model listed above. It should, if possible, reflect all the basic qualities, factors, and relationships that characterize the system, as well as the criteria and restrictions imposed on it. That is, the model should be sufficiently universal to adequately describe the simulated system. On the other hand, to perform the necessary research at a reasonable cost, the model should be relatively simple. These two requirements suggest that computer modeling of systems is an art rather than an established science.

The essence of computer simulation lies in obtaining quantitative and qualitative results for the developed model. Computer modeling is a tool that allows you to obtain quantitative results and make qualitative conclusions that allow one to detect previously unknown properties of a complex system: its structure, development dynamics, stability, integrity, and other.

The process of building computer models can be represented as a sequence of the following steps.

1. Problem statement and its analysis. At this stage, the objectives of the simulation are determined, and the initial information about the system is collected.
2. Construction of a mathematical model. The essential parameters of the model and the relationship between them are determined, and a mathematical description of the dependencies between the parameters is made.
3. Development and implementation of a computer model algorithm. The method of obtaining the initial data is determined, the algorithm is compiled, and its correctness is checked. Next, the tools are selected by the algorithm implementation program, a model is developed, and its correctness is verified.

4. Conducting a computational experiment. A study plan is being developed for the experiment. The analysis of the results is carried out; conclusions are drawn regarding the properties of the model obtained.

In the process of the experiment, it may become clear that it is necessary to:

- adjust study plan;
- choose another method for solving the problem;
- improve the algorithm for obtaining results;
- refine mathematical model;
- make changes to the formulation of the problem.

In this case, the process returns to the appropriate stage and begins again. That is, the process of computer simulation is often iterative.

Currently, there are many software environments for computer modeling, to simplify the development of computer models and conduct computational experiment. Many of these environments allow you to independently create new models from scratch, as well as have extensive libraries of ready-made models (components), on the basis of which you can build other more complex models. It is also usually possible to create your own libraries.

An existing modeling environment is usually based on a modeling language. Modern modeling environments provide features like:

- deriving mathematical equations in the most natural way possible for the user;
- selection of numerical methods for solving algebraic, differential, algebraic–differential, and other equations;
- graphical interface for creating multi-component models;
- interface for setting up and conducting a computational experiment;
- graphic primitives for creating model visualization;
- other opportunities (e.g., conducting an experiment in real time, the possibility of user intervention in the experiment, and others).

Since equations are written in relatively free form, all modeling environments have a tool like a solver. The solver analyzes and transforms the equations to apply numerical methods to them. It is assumed that the user must monitor the fulfillment of the conditions of existence and the uniqueness of the solution.

1.5 Classification of Computer Models

Computer implementations of mathematical models can be classified according to several criteria: by interaction with the outside world, by internal structure, by type of model time, and others.

By the type of interaction with the outside world, models can be divided into open and isolated. The state variables that characterize the model are divided into internal and external. In isolated models, all variables are internal; that is, they are not visible

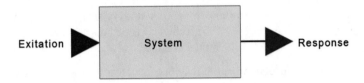

Fig. 1.1 Simplest dynamic model with a response to a given external influence

outside the model. In open models, some of the variables are external, as depicted in Fig. 1.1. They are used to receive and transmit information from the outside, that is, from other components. External variables are of several kinds:

- inputs whose values can only be changed externally;
- outputs whose values vary within the model and can be transferred to other objects (models);
- contacts and streams; their values can change both inside the model and outside.

By the presence or absence of the internal structure of the model are divided into elementary and composite. Elementary models contain only variables and the law of evolution (behavior), and they are also called one component. In turn, composite models can also contain local components, that is, other models, and the connections between them. These models are often called multi-component or simply component.

According to the type of model time, models are continuous, discrete, and hybrid (continuous–discrete). Time can be modeled in different ways depending on the task. As a model of continuous time, the segment of the real axis $[0, T]$ is usually taken, where the value of T is determined by the objectives of the experiment. The time variable t at a given interval varies continuously with a constant speed. The state variables in this case change continuously depending on time, and the model of the dynamic system is continuous, as shown in Fig. 1.2a.

There are, however, models in which it makes sense to monitor only the fact of the occurrence of events, without being interested in the time of their occurrence. Successive events can be renumbered, and we get a sequence of time samples— renumbered clock cycles. In this case, time (t) takes a finite number of integer values equal to the number of ticks in the model. Then, the state variables change discretely, and the system model is called discrete, as shown in Fig. 1.2b.

The combination of segments of continuous time, between which there are discrete cycles, is a model of hybrid time. It is modeled as follows. Before occurrence of some conditions, called an event, time has a continuous behavior, and the variables change continuously. Sections with continuous time are called states. As soon as an event occurs, continuous time "freezes," the local discrete clocks turn on, and the variables change discretely. After all discrete actions have been completed, the discrete time stops, and the continuous time starts to change again (from the point at which it "stopped"). Such events in the model can be as many as one like. Models using this type of time are called hybrid. The states of the hybrid system are described by systems of differential or differential–algebraic equations. Events are defined by logical predicates; discrete actions are correctly described algorithmically. Hybrid

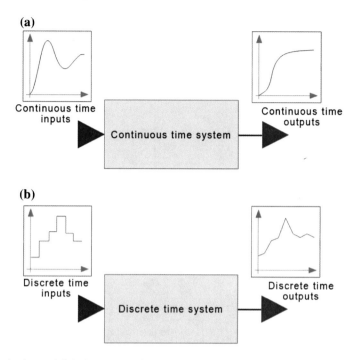

Fig. 1.2 Analog and digital systems. **a** Continuous time system and **b** discrete time system

models allow systems to be described in a variety of engineering applications. For example, in mechanical objects, continuous motion can be interrupted or corrected by some physical impact. In addition, hybrid systems are used in the modeling of heterogeneous systems, of which the elements and subsystems have a different physical nature.

1.6 Open and Isolated Models

As noted in Sect. 1.5, open system models can interact with the environment through external variables, while isolated ones cannot. But on the other hand, when building a model, it is difficult to completely isolate the system from the environment, since the interaction between them still exists. Therefore, as an isolated model of such systems, both the object and the environment in which it is located are considered simultaneously. Thus, an isolated model contains both an object model and an environmental model.

For example, one can consider the free fall of the body from a height h_0 with initial speed v_0, as shown in Fig. 1.3.

Fig. 1.3 Free fall of the
body from a given height

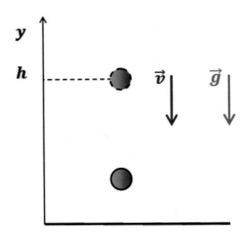

As known, it is described by:

$$\frac{d^2 h}{dh^2} = -g.$$

$$h(0) = h_0; \quad \left.\frac{dh}{dt}\right|_{t=0} = v_0,$$

The parameter g outwardly does not differ in any way from h_0 and v_0, but it is not a characteristic of the object, but of the medium in which it is located, i.e., characterizes the planet on which the experiment is conducted. This means that in reality the object is not isolated, but at the same time, an object model that takes into account the properties of the medium is considered to be isolated.

At the same time, there are cases when it is convenient to consider the object and the environment separately. Then, the model consists of two parts (components): the object model itself and the environment model. The interaction between them occurs through external variables for some specific laws. In this case, they say that the actual model of the object will be open.

Isolated single-component (that is, without internal structure) models are the simplest type of models. All variables of such models are logically divided into state variables, parameters, and constants. The difference between a parameter and a constant is that the constant is set at the stage of building the model and cannot be changed during a computational experiment, whereas the parameters cannot be changed during a separate experiment but can be changed during a series of experiments. It can be said otherwise that the parameters serve to "tune" the properties of the object being modeled.

1.7 Single component and Multicomponent Models

Earlier, we called the object model that does not consider its internal structure as one component (or neutral). In this case, the object can interact with the external environment, which will be considered in the model. If an object selected from the environment is rather complicated and its structure is essential for modeling, then, by creating its model, one can try to reflect its internal structure in the model. In this case, the model is built as a set of objects functioning in parallel and interacting with each other (e.g., a double pendulum). This approach to modeling is called component.

A simulated object can be quite complex itself. For example, a model of a double pendulum made up of sequentially connected ordinary mathematical pendulums can be built in two ways, as shown in Fig. 1.4.

You can consider the double pendulum as an indivisible object, and to obtain the law of evolution, use, for example, the Lagrange equations. In this case, you get a closed system of equations that completely describe the behavior of this object, and it is logical to call it elementary, since it does not consider the structure.

Another way is to split the source object into some elementary components—ordinary mathematical pendulums—and indicate the interaction between them. In this case, the model of a double pendulum will be multi-component; that is, it will consist of several models (components)—ordinary pendulums, each of which considers the effect of the other on itself through forces.

Creating a multi-component model begins with an analysis of the real-world object to highlight the components and the connections between them. When designing from top to bottom, first select the object itself from the environment, and then detail the description of the object as far as necessary, turning it into a multi-component system.

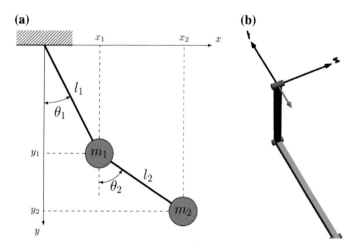

Fig. 1.4 a Graphic representation of a double pendulum for composing Lagrange equations and **b** a component model built in the WSM package from simple mathematical pendulums

Also, component modeling involves the creation of some libraries of elementary components, on the basis of which more complex models can be built. When designing complex objects using libraries of ready-made components, you can move from top to bottom, stopping the detailing process, when the necessary component is found in the library, and from the bottom up, building a model from ready-made components. The lack of a finished component leads to the need to design a new component, which further replenishes the library.

Each component in a multi-component model can in turn be elementary or composite.

1.8 Continuous, Discrete, and Hybrid Models

To better understand time modeling, consider an example of a bouncing ball model depicted in Fig. 1.5. The model should describe the vertical motion of the ball, which falls under the action of gravity and bounces off the floor. The acceleration of gravity is constant and equal to $g = 9.8 \, \text{m/s}^2$. The ball is thrown from a height of h_0 above the floor with a speed of v_0.

The motion occurs in the vertical direction, so the origin of the coordinates will be placed on the floor, and the vertical axis will be directed upward. The variables of the state of the ball will be its height above the floor h and the vertical velocity v. The speed of the ball is positive when it moves up. The following system of two equations describes the free vertical movement of the ball to the first rebound.

$$\begin{cases} \frac{dh}{dt} = v, \\ \frac{dv}{dt} = -g, \end{cases}$$
$$h(0) = h_0; \quad v(0) = v_0,$$

Fig. 1.5 An example of a bouncing ball model

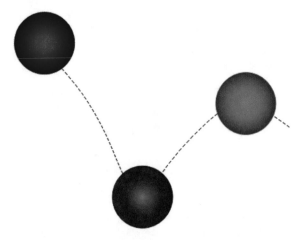

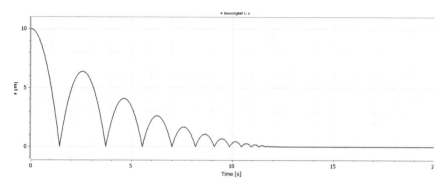

Fig. 1.6 Graph of the position of the bouncing ball above the floor

After a rebound, the ball gains the speed $v_+ = k \cdot v_-$, where v_+ is the speed at which the new flight begins, v_- is the speed at which the ball touched the surface, and k is a coefficient depending on the properties of the rebound. With an absolutely elastic rebound, $k = 1$, with an absolutely inelastic $k = 0$, with an elastic $0 < k < 1$. We are interested in the moments of time at which the bounces occur and the law of change in the height of the ball above the floor, as shown in Fig. 1.6.

When using the model of continuous and discrete time, this problem can be solved only as a sequence of individual tasks. The processes occurring at the time of the rebound can be neglected, that is, assume that the rebound occurs instantly. With this, the rebound phenomenon itself is replaced by the instantaneous action $v_+ = k \cdot v_-$.

The first formulation of the problem: continuous time model.

In this simple problem, the solution of equations for each individual flight (except the first) can be obtained in the explicit form:

$$h(t) = v_0 \cdot t - \frac{g \cdot t^2}{2},$$
$$v(t) = v_0 - g \cdot t,$$

where v_0 is the ball speed after rebound. Then, the next bounce time T can be found from:

$$h(T) = 0 \Rightarrow T = \frac{2 \cdot v_0}{g}.$$

Consistently solving the equations of flight after the next bounce, we obtain the required points in time.

The second formulation of the problem: discrete time model.

We use the solution obtained in order to solve a slightly more "complicated" problem. If we assume that with each rebound $j = 1, 2, 3, \ldots$ the coefficient k_j changes, then

$$T_j = T_{j-1} \cdot k_{j-1}; \quad j = 1, 2, 3, \ldots;$$

$$T_0 = \frac{2 \cdot v_0}{g}; \quad k_0 = 1.$$

Using such difference equations, it is possible to calculate only the moments of time of rebounds, but it is impossible to obtain the law of change of state variables.

The third formulation of the problem: hybrid time.

The question arises: Is it possible to reproduce both long-term and instantaneous processes within one model?

We use the concept of hybrid time. We divide the processes taking place into long-term "flight" and instantaneous—"rebound," and we will alternate long actions with instantaneous ones. Namely, calculate the altitude change (continuous time), suspend lengthy processes upon the occurrence of an event—rebound—perform the specified sequence of instantaneous actions (discrete time), and then continue the long actions.

The ability to alternate between purely continuous and discrete behaviors, clearly separating them, gives undeniable advantages. This allows, within the framework of one model, to combine the incompatible: flight, described by ordinary differential equations, and body rebound from the surface is a task, the necessitating use of the theory of elasticity.

There are three main factors contributing to the emergence of hybrid behavior.

Hybrid behavior due to the combined operation of continuous and discrete objects.

Such hybrid behavior is a characteristic of automatic control systems, in which there are a continuous control object and a discrete control device (controller). The simplest case is a conventional discrete controller, which with a certain tact forms the control action.

For the upper levels of control in complex hierarchical control systems, typical are processes of so-called logical control. In this case, the behavior of the control device is set by an asynchronous process, in which the next discrete event depends in general on the previous one, as well as on continuous variables of the control object. For example, a logic control device for a rocket issues a command to cut off thrust when a certain functional reaches a threshold value.

Hybrid behavior due to instantaneous qualitative changes in a continuous object.

Some systems that are continuous in nature may exhibit discrete behavioral traits associated with the qualitative changes occurring in them. The qualitative changes themselves are primarily due to the multi-mode operation of the system. It should be noted that, unlike the first and third types of hybrid behavior, where hybrid behavior is a natural initial property of the problem itself, the hybrid behavior of this type is artificial to a certain extent and is associated exclusively with the convenient for the researcher formalization of the phenomenon. In fact, discreteness appears here due to the idealization of the initially continuous behavior of real physical systems. The main idealization of this kind is the neglect of the time of transients, when this time is orders of magnitude shorter than the time of operation of the system under study, as well as the idealization of parametric dependencies. An example of the idealization of the first type is the neglect of the time of an absolutely elastic ball rebound from an absolutely solid plane, and an example of the idealization of the second type is the idealization of a real current–voltage characteristic for an ideal diode. In addition, the researcher may simply abstract away from a detailed description of the dynamics of a nonlinear transient process (or a detailed description may simply be unknown), replacing it with some integral dependencies. It should be noted that, due to its artificial nature, hybrid models of this type may exhibit paradoxical behaviors that are not characteristic of the original continuous objects, and the researcher must be extremely careful in formalizing the hybrid model.

Hybrid behavior due to system composition changes.

If continuous objects during operation can appear within the boundaries of the system under study and leave it, then the composition of the total state vector of the entire system x and its dimension will change. Examples of such systems are: an airport (airplanes appear within the airport zone from the outside, land at the airport, and take off from the runway), an air defense complex (targets appear in the detection zone, exit it, and are destroyed by missiles), a system of emerging and disappearing charged particles, etc.

1.9 Linear and Nonlinear Systems

Linear models of dynamic systems are based on systems of linear differential equations. These models are important from the viewpoint of modeling due to the fact that:

- In many cases, linear models are sufficient to reflect the most important properties of the object being modeled, and linear systems are well studied and amenable to qualitative and quantitative analysis.
- In a sufficiently small neighborhood of any point of the solution of a nonlinear equation, one can construct an approximate linear model, the analysis of which makes it possible to judge the local properties of the nonlinear model.

Consider a system of ordinary differential equations:

$$\frac{d\mathbf{x}}{dt} = \mathbf{F}(\mathbf{x}, \mu, t)$$

where t is the time, $\mathbf{x} = (x_1, x_2, \ldots x_n)$ are the time-dependent state variables, μ is the time-independent vector, and $\mathbf{F} = (F, F_2, \ldots F_n)$ are some given functions.

In the general case, the functions $F, F_2, \ldots F_n$ are nonlinear functions of the state variables $x_1, x_2, \ldots x_n$. Such a model is called nonlinear. In the case when all functions $F, F_2, \ldots F_n$ are linear with state variables, then the model is linear and is given by the system of equations

$$\frac{d\mathbf{x}}{dt} = \mathbf{A} \cdot \mathbf{x} + \mathbf{b}$$

where $\mathbf{A}$ is some non-degenerate matrix and $\mathbf{b}$ is the vector of given functions of time.

Linear models can be classified according to different types of behavior of dynamical systems, using information either only on the coefficients of the matrix $\mathbf{A}$ or on its eigenvalues, that is, on the roots of the characteristic equation

$$\det(\mathbf{A} - \lambda \cdot \mathbf{E}) = 0,$$

where $\mathbf{E}$ is the dimension unit matrix $n \times n$.

Singular and fixed points can be found in a known manner [1]. Moreover, each type of a particular point can be associated with a certain phase portrait.

To classify the singular points, one can also use the trace of the matrix $Sp(\mathbf{A})$ and the determinant $\det(\mathbf{A})$.

The solution of linear ODE systems is not difficult for most modern simulation environments. To solve nonlinear systems in modeling packages, as a rule, the possibility of both automatic and manual selections of numerical methods is provided. At the same time, it is recommended that the user himself checks the existence and uniqueness of the solution.

1.10 Component-Oriented Approach in Modeling

As mentioned earlier, component, or multi-component, models, unlike elementary ones, have an internal structure. This means, along with state variables, parameters, constants, and equations linking them, they include local components, which in turn can be both elementary and again component models. Components interact with each other through connections. Modern computer simulation environments provide a graphical interface for component modeling. The components of the model are schematic blocks in a graphic diagram, and the connections between them are defined by lines connecting the corresponding external variables.

The component-oriented approach to modeling is primarily convenient for the user to perceive the model. It also allows you to avoid many errors at the stage of building a model. Naturally, any model can be simply described by a system of equations. However, if the model is complex, such a system may consist of tens and hundreds of equations, and there is always the risk of making a mistake if you compose the system manually. For example, when modeling a complex electrical circuit, it will be necessary, according to Kirchhoff's laws, to describe all the cycles of this circuit, and there may be a lot of them. Component modeling usually involves the creation of libraries of some typical components for a given subject area (current sources, resistors, capacitors, etc.) that can be used in many models. Each component is characterized by its relatively small set of equations, and the aggregate system is compiled automatically.

With this approach, the process of creating a model is reduced to its decomposition, that is, splitting into relatively simple components, which, if necessary, can be refined depending on the task, thus when component modeling models are obtained hierarchical.

As already noted, the components are connected by bonds, which can be of two types: oriented (directed) and non-oriented (non-directional).

This tutorial covers only the basics of the component-oriented approach. More information can be found in [2–4].

Oriented Link Components
Oriented links are used when the direction of information transfer between components of a model is obvious to a model maker.

The most significant example of the use of oriented relations is the automatic control system. In the simplest case, the control system can consist only of a controlled object and a regulator, as shown in Fig. 1.7. The regulator receives the value of the controlled variable x from the control object and, using the difference with the required value x_0 (which can be a parameter and can be supplied externally), calculates the value of the control signal. The control signal is then transmitted to the managed object, and thus the value of the variable x is controlled. All variables are considered dependent on time t.

In this example, the variable x of the managed object is the output, and the variable of the regulator with the same name is the input. Similarly, the variable u in the control object will be input and in the controller output.

Fig. 1.7 Scheme of the simplest model of automatic control system

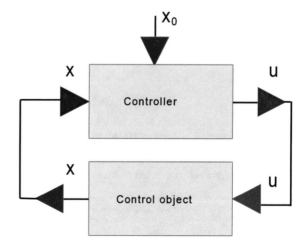

Fig. 1.8 PID regulator
circuit

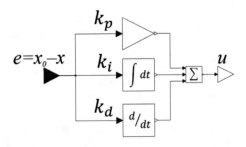

In turn, the model of the regulator can be represented both in the form of an equation and in component form. Suppose that the controller is a proportional-integral-differentiating (PID controller). Then, it can be described by the following system of equations:

$$e(t) = x_0 - x(t),$$

$$u(t) = k_p e(t) + k_i \int_0^t e(\tau)d\tau + k_p \frac{de}{dt},$$

where $e(t)$ is the control error (discrepancy) and k_p, k_i, k_d are the gains of the proportional, integrating, and differentiating components of the regulator, respectively.

The PID controller can also be represented in the form of a flowchart, which is a kind of component model with oriented connections, as depicted in Fig. 1.8.

However, the introduction of a local structure is not always advisable. The representation of the PID controller directly in the form of a system of equations may be more convenient.

From the point of view of the contribution to the total system of equations of the model, each oriented relationship is a formula in which the input variable is assigned a known value of the output variable.

For example, the connection in Fig. 1.9 corresponds to the formula:

$$y = x.$$

It can be interpreted as an equation resolved with respect to a variable y. In other words, the variable y is assigned the value of the variable x.

The connection of inputs and outputs by oriented links is determined by the rules:

- Any input can be connected to several outputs.

Fig. 1.9 Oriented
connection

Fig. 1.10 Non-oriented
relations between external
variables of the contact
(flow) type

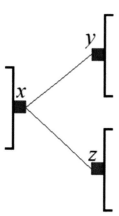

- Any output can be connected to only one input.

In case it is necessary to transfer the values of several variables from one object to another, many computer simulation packages are allowed to combine their logical integer—the record type. This allows you to avoid piles of inputs/outputs and the connections between them in a graphical representation. In this case, the compatibility of the types of bound variables is always checked.

Non-oriented Link Components
Non-oriented links for connecting external variables are used when it is impossible to specify the desired variables in advance; that is, it is impossible to unambiguously indicate the direction of information transfer between components in the model, as shown in Fig. 1.10. In this case, the desired variables are determined after analyzing the total system of equations.

Non-oriented, as a rule, are all physical connections. Systems with such connections include electrical and hydraulic systems. Building models from blocks with non-oriented links are often called "physical" modeling. Typical components are usually used, for which the correctness of component models constructed from them is guaranteed. For hydraulic systems, for example, a set of typical components may be as follows: pressure/flow source, pipe, valve, pump, tank, and others.

External variables that form non-oriented links come in two forms: contacts and flows. In the aggregate system of equations, the "contact" connections correspond to the equations

$$\begin{cases} x - y = 0 \\ x - z = 0 \end{cases}.$$

Any two of these variables may be desired.

Several variables connected by "flow"-type connections correspond to one equation:

$$x + y + z = 0.$$

The desired variable after analyzing the aggregate system will be one of the members of the equation. This equation resembles Kirchhoff's law for electrical circuits.

As an example, consider a closed pipeline circuit with a reservoir. The contour has a section with branching and a pump that starts the circulation of water along the contour, as illustrated in Fig. 1.11.

It is logical to break this system into components and separately describe each component. The main components in hydraulic systems are pipes. The variables connecting the components in this case will be the pressure P and the flow rate Q. Moreover, the pressure will be a variable of the "contact" type, since the pressure at the junction of the two pipes must be equal, and the flow rate will be of the "flow" type to consider the possibility of branching the pipeline; this fluid flow must also be separated. In order not to overload the model diagram, these variables are usually combined into one, declaring the record type.

Thus, it is possible to obtain the scheme of the component model, as shown in Fig. 1.12.

Fig. 1.11 Simple hydraulic circuit

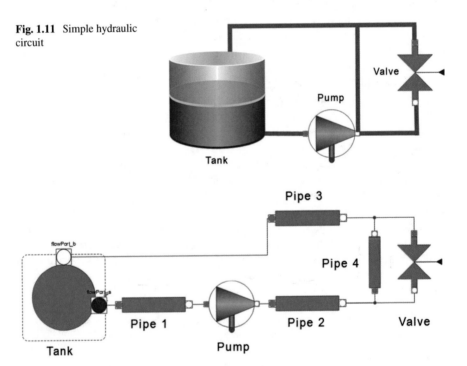

Fig. 1.12 Component model of the hydraulic circuit

This chapter briefly outlines the basic concepts of the theory of modeling. A detailed description of the material can be found in the extensive literature on this subject [1, 5–14].

References

1. A.A. Samarskii, A.P. Mikhailov, Principles of Mathematical modeling: Ideas, Methods, Examples. – CRC Press, Taylor & Francis Group (2002)
2. Yu.B. Kolesov, Yu.B. Senichenkov, Modelirovanie sistem. Ob'ektno-orientirovannyj podhod. – BHV-Peterburg, St. Petersburg (2012)
3. Yu.B. Kolesov, Yu.B. Senichenkov, Matematicheskoe modelirovanie slozhnyh dinamicheskih sistem. – Izdatelstvo SPbPU, St. Petersburg (2018)
4. Yu.B. Senichenkov, Komponentnoe modelirovanie slozhnyh dinamicheskih sistem. Zadachnik. – Izdatelstvo SPbPU, St. Petersburg (2018)
5. N.N. Bautin, E.A. Leontovich, Metody i priyomy kachestvennogo issledovaniya dinamicheskih sistem na ploskosti. – M.: Nauka (1990)
6. K. Velten, Mathematical Modeling and Simulation: Introduction for Scientists and Engineers. – Wiley-VCH Verlag GmbH & Co KGaA (2009)
7. N.V. Butenin, Elementy teorii kolebanij. – L.: Gosudarstvennoe soyuznoe izdatelstvo sudostroitelnoj promyshlennosti (1962)
8. N.V. Butenin, Elements of the Theory of Nonlinear Oscillations. – Blaisdell Publishing Company (1965)
9. Yu.B. Kolesov, Yu.B. Senichenkov, Matematicheskoe modelirovanie gibridnyh dinamicheskih sistem. – Izdatelstvo Politekhnicheskogo universiteta, St. Petersburg (2014)
10. Yu.B. Kolesov, Yu.B. Senichenkov, Modelirovanie sistem. Dinamicheskie i gibridnye sistemy. – BHV-Peterburg, St. Petersburg (2012)
11. E. Haier, S.P. Norsett, G. Wanner, Solving Ordinary Differential Equations I. Nonstiff Problems, 2nd edition, Springer Series in Computational Mathematics. – Springer-Verlag Berlin Heidelberg (1993)
12. Yu.V. Shornikov, D.N. Dostovalov, Osnovy modelirovaniya nepreryvno-sobytijnyh sobytij. – Izdatelstvo NGTU, Novosibirsk (2018)
13. A.A. Yablonskij, S.S. Noreiko, Kurs teorii kolebanij. – M.: Gosudarstvennoe izdatelstvo «Vysshaya shkola» (1961)
14. M. Atansijevic-Kune, S. Blazic, G. Music, B. Zupancic, Control-oriented modeling and simulation: methods and tools. - University of Ljubljana, Ljubljana (2017)
15. V.V. Nemyckij, V.V. Stepanov, Kachestvennaya teoriya differencialnyh uravnenij, izd. 2-e, pererab. i dop. – M.– L.: Gosudarstvennoe izdatelstvo tehniko-teoreticheskoj literatury (1949)

Chapter 2
Description of the Wolfram SystemModeler

2.1 General Concepts

This chapter will allow the reader to briefly get acquainted with the basics of working in Wolfram SystemModeler [1]. You will learn how to use the Model Center to create your own model with components from the Modelica standard library, how to create your own components, and how to run a computational experiment using Simulation Center.

Wolfram SystemModeler is an interactive graphical environment designed for computer modeling of complex multi-level physical and other systems based on the Modelica language and allowing for computational experiments with the models obtained.

Wolfram SystemModeler allows the user to create system models using an extensive library of ready-made physical and logical components written in the Modelica language [2, 3] when creating component models using the drag-and-drop approach.

The basis of creating your own models in the SystemModeler environment is the use of the object-oriented Modelica language.

Modelica is a high-level declarative language for describing mathematical models. It is usually applied to engineering systems and makes it easy to describe the behavior of various types of technical components (e.g., springs, resistors, couplings, etc.). These components can then be combined into subsystems and systems with a very complex architecture.

Modelica is useful in our case for several reasons. First of all, it is technically very convenient. Using complex ready-made algorithms, Modelica compilers allow engineers to focus on the mathematical description of the behavior of high-level components with the possibility of high-performance numerical experiments, without going into solving differential–algebraic equations, performing symbolic manipulations, working numerical solvers, generating code, post-processing, etc.

© Springer Nature Singapore Pte Ltd. 2020
K. Rozhdestvensky et al., *Computer Modeling and Simulation
of Dynamic Systems Using Wolfram SystemModeler*,
https://doi.org/10.1007/978-981-15-2803-3_2

Modelica supports a wide range of mathematical methods that allow one to describe both continuous and discrete behaviors described by hybrid differential–algebraic equations. The language supports both causal (often used for designing control systems) and acausal (often used when creating schema-oriented physical structures) approaches within the same model.

Modelica is a language for modeling the behavior of engineering systems in almost any engineering field. It supports both the creation of physical models and control systems.

SystemModeler consists of a modeling environment—Model Center; runtime computational experiment—Simulation Center; and the Wolfram SystemModeler Link package to communicate with the Mathematica package that connects to SystemModeler.

Getting started with SystemModeler is launching the Model Center. With him, we will start dating.

2.2 Model Center

By default, the Model Center runs with four windows (see Fig. 2.1):

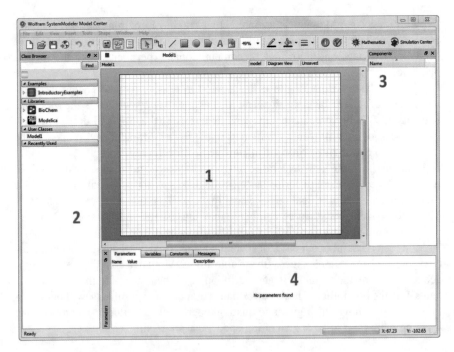

Fig. 2.1 General view of the Model Center

(1) Class Window;
(2) Class Browser;
(3) Component Browser;
(4) other windows (this includes windows for setting variables and parameters, Messages window, and others).

All Model Center windows, with the exception of the Class Window, can be dragged in the main window, as well as moved out of it in the form of floating windows (drag-and-drop principle). When you exit the application, the location and size of the windows are saved. When you reopen the Wolfram SystemModeler, you will see that the window view you have selected is preserved. To reset window settings to "default" view, select "View" -> «Reset Window Layout».

The basis of modeling in the Wolfram SystemModeler environment and the fundamental structural units of modeling in the Modelica language are the components. In the Modelica language, they are referred to as classes, which are elaborated in more detail in the following sections.

Classes (components)

Classes usually contain equations as executable code for computing, but they can also include traditional algorithmic code. All data in the Modelica language is also described by classes. Modelica has the following class types (Table 2.1).

Table 2.1 Types of class

Class type	Description
Model	Designed for acausal, or physical, modeling, i.e., for those systems in which cause–effect relationships are not obvious and may change
Block	Used for causal, or block-based, modeling, i.e., in cases where the causal relationships in the simulated system are obvious. Thus, the difference of the block type is that all its external variables must be either input or output
Connector	Class, which is a special type that serves to communicate between components. The connector cannot have equations and contains only variables involved in connections
Function	A class that describes a function written in an algorithmic language
Package	Packages are used to combine description elements into groups (i.e., class libraries)
Record	Records are used to combine variables of different types into a logical integer. A record may have variables like a model but does not include equations. Records are mainly used to group data. But, they are also very useful in describing data associated with annotations
Type	Class of data type. In addition to using basic types, the user has the ability to declare his own data type using the keyword **type**
Enumeration	Enumerated data type. It is used to define the type of variables that can take only a limited set of specific values

The elements of classes in Modelica can have two levels of visibility: **public** or **protected**. By default, all classes are public. The standard class for describing an executable model is the model **class**.

Class loading

Modelica classes are usually stored one per file (extensions * .mo, * .moe). At any time, you can load one or more Modelica classes into the Model Center by selecting «File» -> «Open». In addition, files can be opened using the drag-and-drop principle; i.e., the required file can be dragged from the explorer to an arbitrary location in the Model Center.

Recently used files are available in the File -> Recent Files menu. To download the desired file, simply click on it. The number of files available in the "Recent Files" menu can be configured in the "General" tab of the "Options" dialog box located in the "Tools" tool menu.

Once the Model Center has finished loading the files, these classes will appear in the class navigator, Fig. 2.2.

Class update

In some rare cases, it may be necessary to force class updates in the Model Center. When a class is updated, the class definition is read from the SystemModeler kernel, and all class members, including child classes, are updated in the Model Center.

To update a class, right-click its name in the Class Browser and select "Refresh" from the pop-up menu, or if the class is open, right-click an empty area in the graphical view and select "Refresh" from the menu.

Fig. 2.2 Recently uploaded and recently used user files

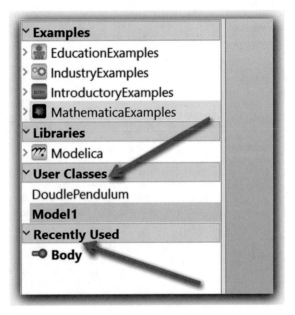

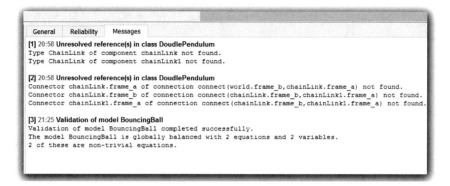

Fig. 2.3 Report on unsuccessful (messages [1] and [2]) and successful (message [3]) model checking

Class check

Verification is done to detect syntactic and semantic errors in the class description. To test the class currently open in the Class Window, click the Validate Class button on the toolbar. You can also right-click on any class in the class navigator and select "Validate" in the menu. Please note that it is impossible to verify classes that have the property "read-only."

 —"Validate Class" button in Tools menu.

As a result of class checking, a message will be generated in the "Messages" window at the bottom of the Model Center about the presence or absence of errors (see Fig. 2.3). Note that semantic checking cannot be performed if the class has syntax errors.

If the test did not reveal any errors, then at the end of the test report there will be information about the total number of equations and variables, as well as the number of non-trivial class equations.

View Classes

The Modelica classes are hierarchically grouped as a tree in the Class Browser, as shown in Fig. 2.4. Groups form packages whose elements are sorted alphabetically. Packages are presented in the form of tree branches. To see the components of a package, expand it by clicking on the arrow symbol to the left of its icon (or name if the package does not have an icon). You can also deploy a package by double-clicking on it. Inside packages, items are ordered according to library developer specifications. The order of the components is specified using the package order file or using the order of declaring files. For more information on packages, see the *Appendix* at the end of this chapter.

You can read more about working with the class navigator in the section on component modeling.

Fig. 2.4 Class
Browser—extensive library
of components Modelica

Appendix: Packages

The **package** *class—a package (class library)—can only contain definitions of classes and constants and serves to implement the principle of encapsulation.*

To combine description elements into groups (that is, class libraries), use packages that are specified by the package class. Packages in Modelica can contain definitions of constants and classes, including all kinds of specialized classes, functions, and subpackages. The term "podpaket" understands that the package is declared inside the other, without implying inheritance relations. Parameters and variables cannot be declared in a package. Definitions in a package usually have to be logically related in some way, which is the main reason for placing them in a package. Packages are useful for several reasons.

Definitions related to a particular topic are usually grouped into a package. This makes finding these definitions easier, and the code becomes clearer.

Packages provide encapsulation and large-scale structuring, which reduces the complexity of large systems. An important example is the use of packages to create hierarchical class libraries.

Name conflicts between definitions in different packages are excluded because the package name is prefixed with the name of the definition declared in the package.

Information hiding and encapsulation can be supported to some extent by declaring protected classes, types, and other definitions that are available only inside the package and therefore not accessible to external code.

Modelica defines a package search method by standard mapping of package names to storage locations, usually in files or directories in the file system.

General package syntax:

```
package PackageName "Description of package"
   // Package contains definitions of classes,
functions and constants.
   end PackageName;
```

Packages can also be nested, for example,

```
    package OuterPackage "Package containing
an external package"
  // Contents of the external package
  package NestedPackage "Internalpackage"
     // Contents of the internal package
  end NestedPackage;
end OuterPackage;
```

More complete information about the packages can be obtained in [2, 4].

Search classes

If you know the class name (or part of its name), you can use the search bar at the top of the class navigator. This function can be useful if you do not know the exact location of the class. After entering the text, all classes corresponding to the input are shown in the search bar, as shown in Fig. 2.5. To see which particular package has the class you need, in the search results you can right-click on it and select "Go to Class in Class Browser" in the pop-up menu.

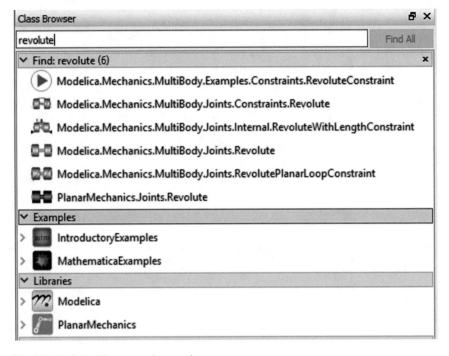

Fig. 2.5 Modelica library search example

Creating classes

To create a new class, on the File menu, click New Class. You can also create a new class by right-clicking on the free space in the class navigator and choosing New Class in the pop-up menu, as shown in Fig. 2.6.

This opens a dialog box where you can set various attributes of the class being created, as shown in Fig. 2.7.

You can set the following attributes:

See Table 2.2.

Modes of operation in the Model Center

The Class Window uses three possible views to display the various properties of the class.

Icon View: graphic mode showing the icon of this class (model). The icon will represent the model we have created in the future if we want to use it as a component of another, more complex model.

Diagram View: graphical mode, showing the chart of this class (model). The diagrams visually show the structure of the model, in the case when components were used to create the model (ready-made or created by the user). Otherwise, the categories window in this mode will remain empty.

Modelica Text View: view mode code written in the Modelica language (textual representation) of the Modelica class.

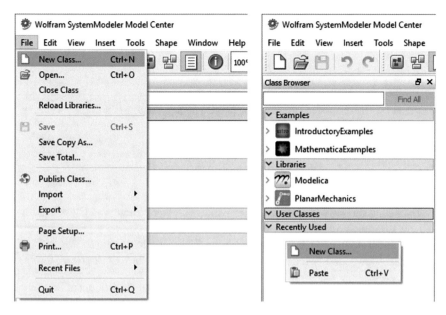

Fig. 2.6 Ways to create a new class: from the File menu (left) and from the class navigator (right)

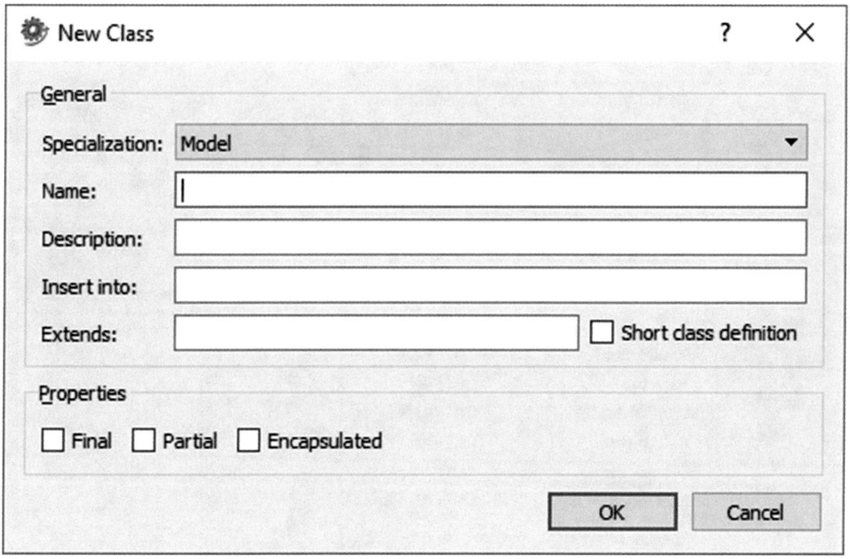

Fig. 2.7 Creating a new class

Table 2.2 List of class attributes

Class attributes	Description
Specialization	Allows you to select a class type
Name	Class name
Description	Text explanation for the class. Optional attribute
Insert into	Allows you to specify the location of the created class to create a hierarchy. This field specifies the name of the class into which the given is inserted into the class tree. Optional attribute
Extends	The field for specifying the parent class, while the class being created inherits the variables and equations from the parent. Optional attribute

To change the view of the categories window, select "View" -> "Class Window" and click on the mode name, as shown in Fig. 2.8.

You can also change the view mode of the Class Window using the "Icon View", "Diagram View", and "Modelica Text View" buttons on the "Class View" toolbar.

—"Icon View", "Diagram View", and "Modelica Text View" buttons on the "Class View" toolbar.

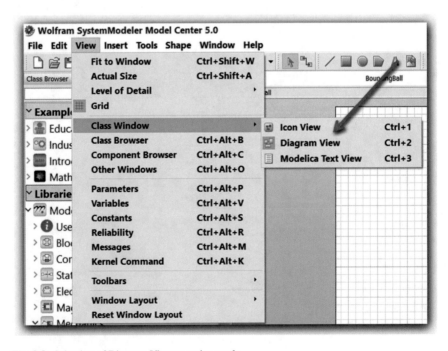

Fig. 2.8 Selection of Diagram View as active mode

Click on the Modelica Text View button on the toolbar to switch to text mode.

Error correction

In case of an error, you can always undo the most recent actions. Choose "Edit" -> "Undo" to undo the last unwanted action you performed. If later you decide that you do not want to cancel a certain action, select "Edit" -> "Redo". Please note that you cannot undo all actions. The Undo and Redo operations can also be found on the standard toolbar.

 Undo and Redo buttons on the standard toolbar.

Set the number of undo levels

The maximum number of consecutive actions that you can undo is 100 by default. You can change this number to any number from 0 to 999. Open the "Options" dialog box by selecting "Tools" -> "Options". This setting is available in Model Center ▶ General and is automatically saved when you close the dialog box and is restored the next time you run Model Center.

At the end of this section, it is worth saying that if you still have any questions, you can always use the prompts.

Tooltips

Tips are small rectangles with auxiliary messages that appear when you hold the cursor over an element of interest, or, for example, a button on the toolbar, or a connection (connector) in the Diagram View viewing mode, as shown in Fig. 2.9.

In some modes, such as the Diagram View viewing mode, tooltips can be disabled. To enable or disable tooltips for a specific view, open the "Options" dialog box by selecting "Tools" -> "Options". Tooltip settings are located in Model Center -> General and are automatically saved when the dialog box is closed and restored the next time you run Model Center.

In most dialog boxes, you can get a hint to any item in the dialog box, by clicking the question icon in the title bar, then clicking the item or right-clicking the area or item, and choosing "What's This?" from the pop-up menu.

You will learn more about creating a model using ready-made or created classes in Sect. 2.4.

Fig. 2.9 Tooltip is indicated by a red arrow

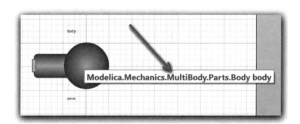

2.3 Simulation Center

A numerical experiment is performed using the Simulation Center. To start a numerical experiment, click the "Simulate Class" button on the "Tools" toolbar. This will launch your model on the numerical experiment in the Simulation Center.

▶ ▾ "Simulate Class" button in Tools menu.

By clicking the arrow next to the Simulate Class button, you can set experiment parameters, such as the start and end time of an experiment for a given model.

Simulation Center can also be launched or made visible without creating an experiment by clicking the "Simulation Center" button on the "Applications" toolbar.

Simulation Center   "Simulation Center" button in "Applications" menu.

After switching to the execution environment of the Simulation Center, you will see the display similar to that shown in Fig. 2.10. By default, the Simulation Center runs with three windows:

(1) Experiment Browser;
(2) Graphing window;

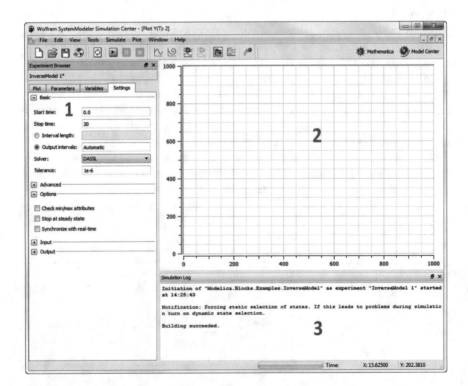

Fig. 2.10 General view of the Simulation Center

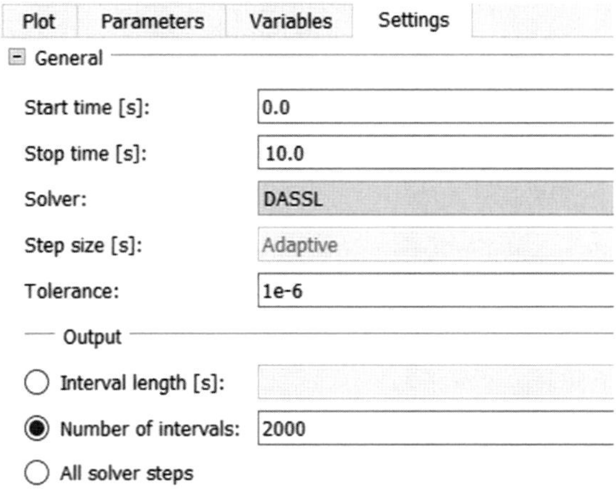

Fig. 2.11 Default
Experiment Browser settings

| Plot | Parameters | Variables | Settings |

☐ General

Start time [s]:	0.0
Stop time [s]:	10.0
Solver:	DASSL
Step size [s]:	Adaptive
Tolerance:	1e-6

Output

○ Interval length [s]:

◉ Number of intervals: 2000

○ All solver steps

(3) Simulation Log.

The "Experiment Browser" and "Simulation Log" windows can be dragged anywhere in the main window of the Simulation Center, or you can create floating windows. The graphing window can be dragged into the main window, but it cannot be made floating.

If there are errors in the report model, then there is an error message in the report model. For more information about the Simulation Log report window, see reference [1].

If the graph window is open, it will be regularly updated during the experiment with the values of the selected variables in the Experiment Browser. This allows you to see intermediate simulation results until it is complete.

If the model does not contain annotations relating to the experiment, all new experiments when they are created will go with the default settings of the experiment, as shown in Fig. 2.11.

2.3.1 Experiment Browser

Experiment Browser is used to view the characteristics of the experiment and their changes. The window of any experiment has four tabs for different characteristics:

- **Plot view**—Here, you can choose which variables and parameters to plot on the chart.
- **Parameters view**—Here, you can edit the values of the parameters of the experiment.
- **Variables view**—Here, you can set the initial values for state variables.
- **Settings view**—It contains all the settings for the experiment.

To change the browser view, click the appropriate tab at the top of the browser.

The installation of parameters and initial values, as well as the choice of the variables of interest to us to build graphical dependencies, is quite obvious. If difficulties arise, it is recommended to refer to [4].

Let us dwell in more detail on the choice of basic experiment settings (on the choice of additional "advanced" settings, see [4]).

To perform a computational experiment, it is necessary to choose a numerical method by which the equations determining the mathematical model of the dynamical system under consideration will be solved. It is also necessary to configure the following experimental parameters (Fig. 2.12):

- start time;
- stop time;
- step size;
- tolerance.

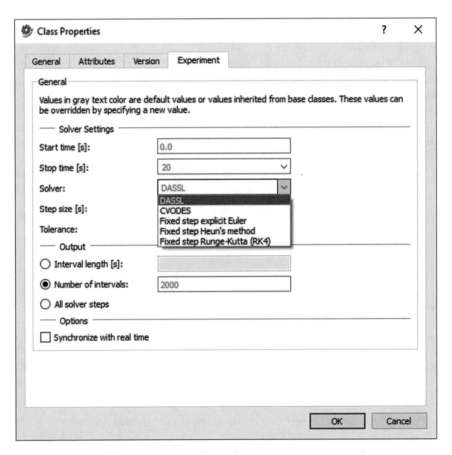

Fig. 2.12 Computational experiment settings

In addition, you should set the number (number of intervals) or the length of the intervals (interval length) to display graphical results, although the listed parameters can also be taken as specified in WSM by default.

2.3.2 Solver Selection

In most cases, the mathematical model of dynamic systems is determined by differential (ordinary) and algebraic equations. There is an extensive literature on the theory and methods for solving these equations [5–22].

To solve systems of algebraic and differential equations specifically in the Wolfram SystemModeler environment (version 4.2) [1], the following methods are used (solver field, Fig. 2.12):

- DASSL;
- CVODES;
- explicit Euler method (with fixed step);
- Heun's method (with fixed step);
- fourth-order Runge–Kutta method (with a fixed step).

In this case, all numerical methods in this version of Wolfram SystemModeler allow solving only ordinary differential equations of the first order. In this regard, ordinary differential equations that determine the mathematical model of the dynamical system under consideration should be reduced to first-order differential equations.

In the settings of the computational experiment, it is possible to independently choose the method by which the determining equations are included in the mathematical model, as shown in Fig. 2.12.

The DASSL and CVODES methods themselves control the integration step, whereas for the rest of the methods the integration step is fixed and can be set by the user (the step size field).

Since the number of integration steps is usually large, not all steps can be displayed on charts, but only a certain number of them (the number of intervals field). Or, alternatively, you can specify the step length (the interval length field) with which points will be drawn on the graphs. Such settings can be useful, for example, if the function oscillates quickly and its graph does not look smooth enough.

DASSL

Differential/Algebraic System Solver (DASSL) [19] is designed for the numerical solution of implicit systems of differential or differential–algebraic equations for $y \in \mathbb{R}^N$ with given initial conditions:

$$F(t, y, y') = 0, \quad y(t_0) = y_0.$$

DASSL is useful for solving two classes of problems that cannot be solved by standard methods. The first class is represented by tasks that cannot be brought to a standard form $y' = f(t, y)$. The tasks of the second class are theoretically reducible to the standard form, but such a transformation leads to serious problems. For example, conversion $Ay' = By$ requires multiplication by an inverse matrix A^{-1}. If matrix A is sparse, then the matrix A^{-1} will no longer exist, so the solution to the original problem is more preferable. In particular, DASSL allows solving differential–algebraic equations when the matrix $\partial F / \partial y'$ is degenerate.

DASSL is based on the backward differentiation formula (BDF), which is part of a family of implicit multi-step numerical integration methods [13, 18]. At each time step, left finite differences (backward differences) are used to approximate time derivatives (backward differentiation):

$$y'(t_n) \approx \frac{y(t_n) - y(t_{n-1})}{\Delta t_n} \approx \frac{y_n - y_{n-1}}{\Delta t_n}.$$

Hereinafter, y_n is an approximation of the exact value $y(t_n)$ and $\Delta t_n = t_n - t_{n-1}$ is a variable (in the general case) time step. Nonlinear system obtained at each time step

$$F\left(t_n, y_n, \frac{y_n - y_{n-1}}{\Delta t_n}\right) = 0$$

solved by Newton's method:

$$y_n^{m+1} = y_n^m - \left(\frac{\partial F}{\partial y'} + \frac{y_n - y_{n-1}}{\Delta t_n}\frac{\partial F}{\partial y}\right)^{-1} F\left(t_n, y_n^m, \frac{y_n^m - y_{n-1}}{\Delta t_n}\right),$$

where y_n^m is the approximation y_n obtained at step m. To accelerate convergence, the nonlinear system is written as

$$F(t, y, \hat{a}y + \beta) = 0,$$

where $\hat{a}$ is the constant characterizing the change in step and β is the vector, depending on the solution at the previous step. The values t, y, $\hat{a}$, and β are calculated at $t = t_n$. The converted system is solved by the modified Newton method.

A more detailed description of the software is presented in [13]. Since DASSL is quite stable for a wide range of tasks, in Wolfram SystemModeler it is chosen by default.

CVODES
C-language variable-coefficients ODE solver (CVODES) [21, 22] is a software tool for solving the Cauchy problem for systems of ordinary differential equations of the first order (both rigid and non-rigid) with respect to $y \in \mathbb{R}^N$, specified explicitly

$$y' = f(t, y), \quad y(t_0) = y_0.$$

CVODES uses one of two numerical methods: using the difference approximation back method of the inverse differentiation (BDF) [13] and the implicit Adams–Moultonc method [11, 12]. Both methods can be represented by a linear multi-step formula.

$$\sum_{i=0}^{K_1} \alpha_{n,i} y_{n-i} + \Delta t_n \sum_{i=0}^{K_2} \beta_{n,i} y'_{n-i} = 0,$$

where $y'_n = f(t_n, y_n)$. The coefficients $\alpha_{n,i}$ and $\beta_{n,i}$ uniquely determine the specific integration formula, with $\alpha_{n,0} = -1$.

Adams–Fulton method is used for non-rigid tasks. He is answered with $K_1 = 1$ and $K_2 = q - 1$. The order of the formula q can vary from 1 to 12. For hard problems, BDF is applied with $K_1 = q$ and $K_2 = 0$. The order of the formula q can vary from 1 to 5, and it is selected automatically and dynamically.

In both cases, a system of nonlinear equations is solved for each value t_n

$$G(y_n) \equiv y_n - \Delta t_n \beta_{n,0} f(t_n, y_n) - a_n = 0,$$

$$a_n = \sum_{i>0} \left(\alpha_{n,i} y_{n-i} + \Delta t_n \beta_{n,i} y'_{n-i} \right).$$

For non-rigid tasks, the iterative process used does not require solving systems of linear equations. For hard problems, the Newton method is used, which requires solving at each step a system of linear equations of the form

$$M\left(y_n^{m+1} - y_n^m\right) = -G(y_n^m)$$

where y_n^m is the approximation obtained at step m and M is the approximation of the Jacobi matrix:

$$M \approx I - \gamma J, \quad J = \partial f / \partial y, \quad \gamma = \Delta t_n \beta_{n,0}$$

If the right side and/or the initial conditions of the problem depend on the parameter $p \in \mathbb{R}^M$:

$$y' = f(t, y, p), \quad y(t_0) = y_0(p),$$

CVODES can also perform forward or reverse (adjoint) sensitivity analysis. A more detailed description of the software is presented in [21, 22].

Explicit Euler Method

The Euler method is the simplest method [11, 12] of solving the Cauchy problem for first-order differential equations given explicitly:

$$y' = f(t, y), \quad y(t_0) = y_0.$$

This explicit one-step method of the first order of accuracy is based on an approximation by an integral curve by a piecewise linear function (Euler broken line), each step of which has the form:

$$y_n = y_{n-1} + hf(t_{n-1}, y_{n-1}).$$

Hereinafter, $\Delta t = h$ is a constant grid step in time. An analysis of the stability of various variants of the Euler method is given in [17].

Heun's method

Heun's method [19] is a particular variant of the general scheme of the Runge–Kutta method, which has a second order of accuracy. It is designed to solve differential equations of the first order, given explicitly

$$y' = f(t, y), \quad y(t_0) = y_0.$$

The method involves a double calculation of the value of the right side of the equation at each step t_n: predictor

$$\hat{y}_n = y_{n-1} + hf(t_{n-1}, y_{n-1})$$

and corrector

$$\hat{y}_n = y_{n-1} + \frac{h}{2}\big(f(t_{n-1}, y_{n-1}) + f(t_n, \hat{y}_n)\big).$$

The analysis of the stability of the method and its properties is presented in [14, 17, 19].

Fourth-order Runge–Kutta method

Runge–Kutta methods are a large class of numerical methods for solving the Cauchy problem for first-order differential equations specified explicitly:

$$y' = f(t, y), \quad y(t_0) = y_0.$$

A variant of the fourth-order Runge–Kutta method [11, 14] implies a fourfold calculation of the value of the right side of the equation at each step t_n:

$$\begin{cases} k_1 = f(t_{n-1}, y_{n-1}) \\ k_2 = f(t_{n-1} + \frac{h}{2}, y_{n-1} + \frac{k_1}{2}) \\ k_3 = f(t_{n-1} + \frac{h}{2}, y_{n-1} + \frac{k_2}{2}) \\ k_4 = f(t_{n-1} + h, y_{n-1} + k_3) \end{cases}, \quad y_n = y_{n-1} + \frac{h}{6}(k_1 + 2k_2 + 2k_3 + k_4).$$

The analysis of the stability of the method and its properties is presented in [12, 14].

2.3.3 Plotting

After you have prepared the initial data for the experiment, the "Plot" tab contains all the necessary variables and parameters of the experiment. They are presented in the form of a tree, which is built hierarchically with components in the form of branches.

Double-click the component name to expand or collapse its branch. By expanding a branch, you can access the variables and parameters of the corresponding component. Each variable and parameter have a checkbox for selecting this element for plotting.

There are two different graphing modes:

- the plotting mode $Y(T)$, which is used to plot the dependence of variables and parameters on time (normal time schedules);
- the plotting mode $Y(X)$, which is used to plot the dependence of a variable or parameter on another variable or parameter (so-called phase diagrams).

To begin with, we will construct the schedule of the time chart. To do this, click on the button on the toolbar ⋀ *New Y(T) Plot Window*, as shown in Fig. 2.13.

In the Experiment Browser window, check the box, for example, opposite the variable x to plot the dependence of the variable x on time, as shown in Fig. 2.14.

An example illustrating the dependence of the displacement angle on time for a mathematical pendulum is shown in Fig. 2.15.

Fig. 2.13 Selection of the type of graph showing the time dependence

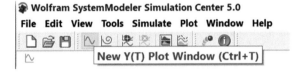

Fig. 2.14 Checkbox for plotting the variable x versus time

Experiment Browser

DiffEq 1*

| Plot | Parameters | Variables | Settings |

⊞ Filter
⊟ Suggested Plot Variables
der(x)

Name	Unit	Description
☐ der(x)		Derivative of x
☑ x		

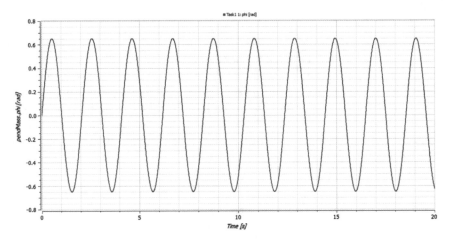

Fig. 2.15 Graph of the offset angle of the mathematical pendulum from time ($\omega = 2$ rad/s) with the given initial conditions

In some cases, it is useful to consider the motion of the system in phase space. The phase diagram method is convenient for the qualitative analysis of dynamic systems. In any system with one degree of freedom, the offset and speed change with time. The state of the system at each time point can be characterized by two values of x and υ, and on the plane of these variables this state is uniquely determined by the position of the imaging point with coordinates (x, υ). Over time, the imaging point will move along a curve, which is called the phase trajectory of motion. Analysis of the trajectory allows you to judge the features of the process. The plane of the variables x and υ is called the phase plane. The family of phase trajectories forms a phase portrait of a dynamic system.

Construct a phase diagram.

To do this, click on the $\boxed{\circ}$ *New Y(X) Plot Window* button in the toolbar, Fig. 2.16.

Next, in the Experiment Browser window, first check the box next to the variable that will be plotted along the abscissa and then the box next to the variable postponed along the y-axis. For example, to obtain the dependence of the angular velocity on the angle of displacement in the problem of the mathematical pendulum (Sect. 2.3.1), it is necessary to set the flags as shown in Fig. 2.17.

As a result, you will obtain a graph as presented in Fig. 2.18.

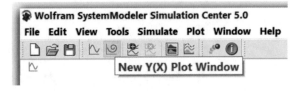

Fig. 2.16 Selecting the phase diagram building mode

☑ phi	°	angle
☐ r[1]	m	**transl. position**
☐ r[2]	m	**transl. position**
☐ specular...		**Reflection of ambient light (= 0: light is completely absorbed)**
☐ sphereD...	m	**Diameter of sphere**
☐ v[1]	m/s	**velocity**
☐ v[2]	m/s	**velocity**
☑ w	ra...	**angular velocity**
☐ z	ra...	**angular acceleration**
☐ zPosition	m	**z position of the body**

Fig. 2.17 Window Experiment Browser

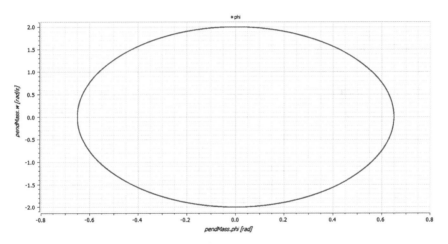

Fig. 2.18 Phase diagram of the mathematical pendulum ($\omega_0 = 2$ rad/s) with a given initial conditions

If you need to repeat a numerical experiment with other values of parameters within one task, you need to perform the following steps:

- To create a new graph, click the icon ⟨icon⟩ again or ⟨icon⟩ on the toolbar.
- To display a variable on a chart, put a tick in the "Plot" window (to the left in the Simulation Center window).
- To create a new graph in the same window, next to the original one, click the icon ⟨icon⟩ on the toolbar.
- To change the value of a parameter, first click "Parameters" next to the "Plot" tab, expand the tree in the "Parameters" view, and enter a new value for the parameter.

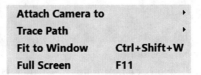

Fig. 2.19 Animation settings window

- By running the simulation, you get the results of a numerical experiment with new values of parameters.

2.3.4   Model Animation

In some cases, when using ready-made components (classes) you will be able to see the 3D animation of your model. To do this, when the numerical experiment is performed, click the Animate ⚙ button in the Simulation Center window.

Use the mouse to rotate the animation at the angle you want to view and zoom in using the scroll wheel. Then, click on the Play button ▶ to play the animation.

Clicking on an empty area of the screen with the right mouse button will result in a pop-up window with additional features, allowing you to look at the animation from different points, increase and decrease the screen size, and trace the path of the selected element, as shown in Fig. 2.19.

Select the trace path function "Trace Path", and check the items you want to trace. Figure 2.20 shows an example of animation of a connected pendulum.

2.3.5   FFT Analysis

To find the natural frequencies of oscillatory systems, it is convenient to use the Fourier transform to go from the timescale to the frequency scale. To do this, you need to select the FFT analysis tool in the toolbar, as shown in Fig. 2.21.

You will see a pop-up window for FFT analysis, in which you need to set the start and end times of the experiment and the limits of the frequencies you are interested in. Remaining values leave "by default". In addition, it is necessary to determine the amplitude of the bias you are interested in.

If the body makes an oscillatory motion, which is described by the displacement of $r[1]$ along the x-axis, then drag the variable $r[1]$ into the FFT analysis window, and then click OK with the set values, as shown in Fig. 2.22.

In Fig. 2.23, the result of the FFT analysis of the x-coordinate of the body oscillat-

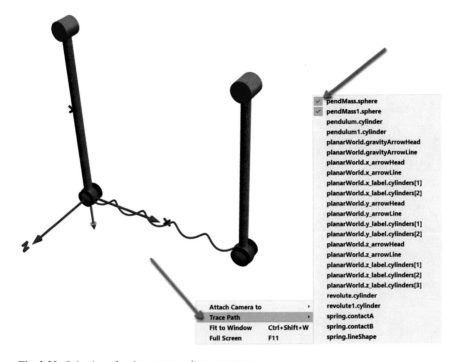

Fig. 2.20 Selection of an item to trace its movement

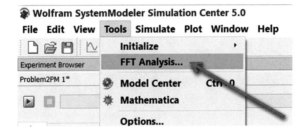

Fig. 2.21 FFT analysis tool selection

ing on the spring in the system of coupled pendulums for a given frequency change from 0 to 2.5 Hz is presented.

Fig. 2.22 FFT analysis tool settings window

FFT Analysis   ?   ✕

Input

Experiment: Problem2PM 1

Variable: pendMass.r[1]

Time start: 0

Time end: 10

Options

☑ Remove DC component

☐ Sampling Interval

4.9975E-3

☐ Number of FFT Points

2002

Cut-off frequency [Hz]: 100.05

OK   Cancel

Fig. 2.23 FFT analysis result

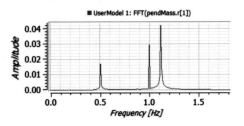

2.4 Creating a Dynamic Model in Wolfram SystemModeler

2.4.1 Creating a Component in Text Mode

Now let's try using the simplest example to go all the way from creating a model to simulating it and creating an icon for further use. Select the New Class command from the File menu, as shown in Fig. 2.24.

Fig. 2.24 Selecting the **New Class** command from the **File** menu

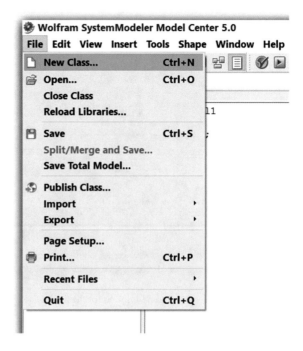

This will open the New Class dialog box in which we will enter the name and description for the class being created. We use one of the base class types. In Modelica, there are two such types: **model** and **block**.

The **model** and **block** classes are the basic standard types of classes that are used to describe various models and their components and which may contain parameters, variables, connectors, local components, equations (including link equations), algorithms. The difference between these types is that they are designed to implement different types of modeling. The **block** type is used for *causal*, or *block-oriented* modeling, i.e., in cases where the causal relationships in the simulated system are obvious. The **model** type is intended for *acausal* or *physical* modeling, i.e., for those systems in which cause–effect relationships are not obvious and can change. Thus, the difference between the **block** types is that all its external variables must be either input or output.

For the first example, the **model** type is appropriate.

Set the class type (Specialization) **model,** and give the created model the name (Name) "**DiffEq**" and description (Description) "**A differential equation**", as shown in Fig. 2.25.

The fact is that in the following, when describing mechanical models, we will all the time deal with ordinary differential equations. Therefore, it makes sense to immediately consider such an example and create the corresponding own component.

After clicking on the OK button, the model will be created and will appear in the Class Browser window. At the same time, the model will open in the Class Window.

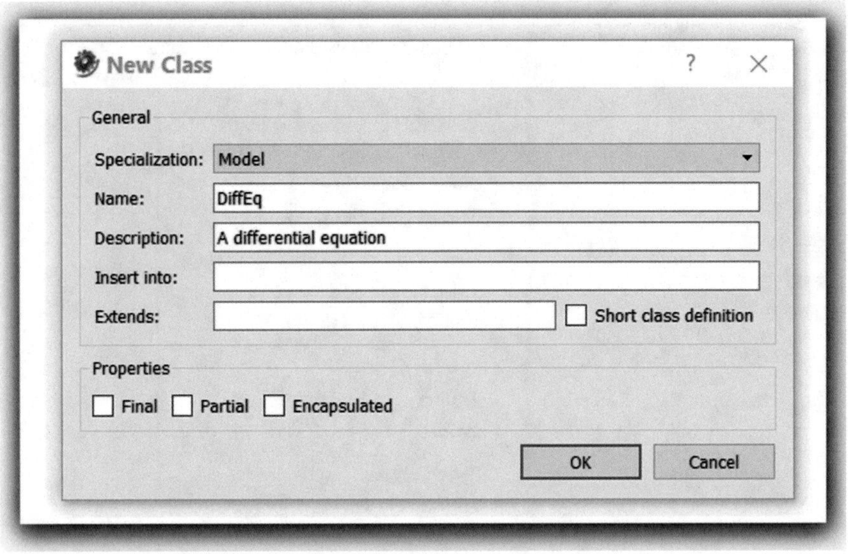

Fig. 2.25 Enter the name and description for the new model

In this case, you must go to the text mode. Consider the rules for creating a model in text mode. You will see that the following lines were automatically created on the sheet, as shown in Fig. 2.26.

Notice the description we entered in the dialog box that was added to the model.

The end of the model is indicated by the keyword end, followed by the repetition of the model name. Any text appearing after the sequence // to the end of the line or between the delimiters / * and * / is considered a comment.

The ¤ symbol in the second line contains hidden graphical information about the model and is automatically updated each time you edit the model in any of the graphical modes of the categories window.

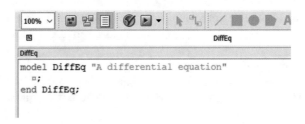

Fig. 2.26 Text View of the model being created

If this model is a descendant of another class, then this fact is passed by the keyword **extends**, followed by the name of the base class. Keep in mind that Modelica supports multiple inheritances. In this case, inheritance can be performed with modification of the parameters of the base classes.

Thus, the definition of the model has the following general form:

```
model ModelName "An optional description"
    extends Model1; // Inheritance without
modifications
    extends Model2(n=5);// Inheritance with
modification
    // In this block, variables, parameters and
constants are declared.
  equation
    /* This block describes the equations. */
  end ModelName;
```

More information about the description of models in the Modelica language can be found in [4].

Now, it is time to add a variable and an equation.

Consider setting variables. The basic data types in Modelica are represented by the following types:

Real—real data type;
Integer—integer data type;
String—string data type;
Boolean—Boolean data type.

They are also classes and possess their properties, for example, can act as base classes. For variables of type **Real**, the function **der()** is defined, which returns the value of the time derivative.

Addition: additional data types

Record

Record type is a class type that serves to combine into a logical integer variable of different types. A record may have variables like a model but does not include equations. Records are mainly used to group data. But, they are also very useful in describing data associated with annotations.

Record looks like a model, but without any equations:

```
record RecordName "Description of the record"
  // Variable declarations
end RecordName;
```

*The entry description begins and ends with the name of the entry being defined.
An explanation of the entry may be included in the line after the name. All variables
associated with the entry are declared in the entry definition.*

The following are examples of entries:

```
record Vector "Vector of three-dimensional space"
  Real x;
  Real y;
  Real z;
end Vector;
```

```
record Complex "Complex number"
  Real re "Real part";
  Real im "Imaginary part";
end Complex;
```

Some details about the records and their creation can be found in [4].

*In addition to using base types, the user has the ability to declare his own data type
using the type keyword. In this way, you can configure specific attributes, for example,
maximum/minimum value, units of measure, and others. Syntactically setting a new
type is as follows:*

```
    type NewTypeName = BaseTypeName
(/* redefinable attributes*/);
```

*Most often, one of the built-in types (e.g., Real) is used as BaseTypeName. But
it may be another derived type. In this way, several levels of specialization can be
maintained, for example,*

```
    type Temperature = Real(unit="K");// Can be
used for temperature difference.
    type AbsoluteTemperature = Temperature(min=0);
// Must be positive
```

In addition to the above types, Modelica also has an enumeration type—enumerations. It is used to determine the type of variables that can take on only a limited set of specific values. In fact, enumerations are not strictly necessary in the language. Their values can always be represented by integers. The enumeration type is specified using the **enumeration** keyword.

Also, to create new data types can be used **record**, which will be discussed below. More information about data types in Modelica can be found in [4].

Let's go to the basic equation.

Equation

Consider the types of equations that we can use when creating a model. First of all, these are ordinary equations.

All equations consist of left and right sides, separated by an equal sign, i.e., representable as:

```
    <left-handside> = <right-hand side>
"Description of the equation (abstract)";
```

Newton's law can serve as an example of an equation in mechanics:

```
m*der(v) = F "Newton's Second Law";
```

The left and right sides of the equation in Modelica are expressions, not assignments. In other words (and unlike most programming languages), the left side should not be variable (as we see in the case of Newton's law above).

Another type of equation is the *initial equation*.

You can specify the equations in the model that will be used to set the initial conditions as follows:

```
    initial equation
x = 0; // Only used to solve for initial conditions
```

Finally, *conditional equations* can be used.

Below we discuss how to use **if** statements to represent conditional behavior. There are two forms of conditional equations. The first is a balanced form, for example,

```
if  a>b  then
    x = 5*time;
else
    x = 3*time;

end if;
```

In the balanced case, the number of equations is always the same (one in the code above), but each equation can change. This is important because for modeling in Modelica the number of variables must be equal to the number of equations, and the number of equations must be constant during the simulation.

Another type of conditional equations is equations in which the number of equations is not balanced. This means that the number of equations on the **if** side may not be equal to the number of equations on the **else** side (as was the case in the balanced case earlier), for example,

```
. .
    parameter Boolean steady_state;
initial equation
    if steady_state then
        der(x) = 0;
        der(y) = 0;
    end if;
. .
```

In other words, if the logical parameter **steady_state** is **true**, then the original equations will be satisfied. But if the parameter is **false**, it is not. The conditional expression here explicitly has parametric variability, because the expression contains only a variable, and this variable is a parameter. As a rule, the unbalanced form of the if operator is used in the initial equations.

Basic information about the types of equations is contained in [4].

Consider the simplest differential equation:

$$\frac{\mathrm{d}x}{\mathrm{d}t} = -x$$

With a given initial condition:

$$x_0 = 1$$

We write the equation with this initial condition in the Modelica language in the text form Text View, as shown in Fig. 2.27.

Fig. 2.27 Text View of the finished model

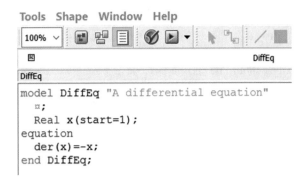

Notice that when we declare a variable, we set its initial value to 1 on the same line, by setting the value for its **start** attribute.

We will discuss in more detail the task of initial conditions in the Modelica language.

In order to solve differential equations, initial conditions must be specified, or initial values for variables. In Modelica, there are several ways to set initial values for variables. Consider the most common.

Initial values for variables can be set either by using the **start** attribute or by adding equations to the section of initial equations/algorithms (how the compiler chooses state variables depends on several factors; for example, variables that are inside the **der** (derivative) operator are preferred. Control over state switching is provided by variable attributes such as **stateSelect** and **fixed**).

If a variable is a state variable, setting its initial value is often crucially needed to ensure that it is appropriate for the start of a computational experiment. By default, the initial value is 0 for numeric (integer and real) variables, takes the value false (false) for logical variables, and takes the form of an empty string "" for a text variable.

While state variables at the beginning of a computational experiment will always have the value specified by the **start** attribute, those variables that are not state variables may not have such attributes. For variables that are not state variables, the value of the **start** attribute is treated as an expected value. In the example below, x is a state variable (as used in an expression with a derivative), and its initial value is set to 1, so at the beginning of the computational experiment, the value of x will be 1.

```
model Model1
Real x(start=1);
equation
 der(x)  = -x + 1;
end Model1;
```

The **start** attribute can also be used to help the resolver find the correct value for a variable that is not a state variable. This can be explained in the example below.

The equation $z^2 = 6 - z$ has two solutions, 2 and -3, respectively. To help the solver find the desired solution, an initial (estimated) value can be given for z. This value should be chosen as close as possible to the real solution. For example, if the initial value 4 is used for z, the solution will be $z = 2$. If the initial value is -5, then the solution will be $z = -3$. See **Model2** and **Model3** below.

```
model Model2
Real z(start=4);
equation
z^2=6-z;
end Model2;

Result: z=2
```

```
model Model3
Real z(start=-5);
equation
z^2=6-z;
end Model3;

Result: z=-3
```

The estimated values are useful when working with nonlinear equations, in particular, if the problem is to start the simulation from a relatively stable state. In this case, the initial values for state variables can be used as estimated values.

It should be borne in mind that the initial value specified by the **start** attribute is only recommended for the solver. If there are a large number of variables and equations in the system, the solver selects initial values for the variables so that they agree with each other; therefore, the actual initial values will not always coincide with the values specified by the **start** attribute. If it is necessary to "hard" set initial values, the attribute **fixed=true** is used (default is **false**) or other methods given below. However, it is recommended to use the **fixed** attribute with caution, since a large number of "hard" given initial values can contradict each other, which will make it impossible to perform a computational experiment.

Another method of defining initial values for variables is to use sections of **initial equations** or **initial algorithms**, which can be used, for example, to specify initial values of derivatives (see **Model4**).

In **Model4**, x will always begin with a value of 1, while in **Model1**, the initial value of the variable x can be easily changed by editing. If the value of a variable is given by the initial equation, it is impossible to change its value by simply editing the initial attribute. The initial equation section is more influential than the initial attribute in all cases where the initial value can be calculated using the equation. For example, in **Model5**, the value of the variable x will start from a static state, since **der(x)=0** is used as the initial equation.

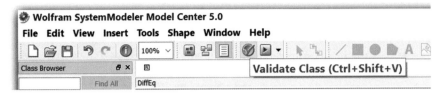

Fig. 2.28 Validate Class button of Model Center toolbar

×	General	Messages

[1] 20:30 **Validation of model DiffEq**
Validation of model DiffEq completed successfully.
The model DiffEq is globally balanced with 1 equation and 1 variable.
1 of these is a non-trivial equation.

Fig. 2.29 **Messages** tab of the **Model Center** message box

```
model Model4
  Real x;
initial algorithm
 x:=1;
equation
  der(x) = -x + 1;
end Model4;
```

```
model Model5
  Real x;
  Real y(start=3);
initial equation
  der(x)=0;
equation
  der(x) = -x + y + 1;
  der(y)= -y +2;
end Model5;
```

More information about the initial conditions can be found in [4].

Now, the **DiffEq** model is ready. Before running the model on the experiment, we may want to perform its verification. Click the **Validate Class** button, Fig. 2.28.

This will create a report in the message window on the Messages tab located below the Class Window.

If everything was entered correctly, you should receive a report similar to that shown in Fig. 2.29.

2.4.2 Perform a Numerical Experiment

When you start Simulation Center, the **DiffEq** model will be automatically converted to an executable program. The experiment for the **DiffEq** model is created in the Experiment Browser window. The default settings are suggested, as shown in Fig. 2.11.

In the Experiment Browser window, you can specify simulation settings, parameter values, as well as initial values for variables. But we will leave the default settings for now by clicking the Simulate button instead to start the simulation.

After performing the simulation of our model, you need to go to the chart view on the Plot tab.

Now, we are interested in time dependencies, as shown in Fig. 2.30.

Check the box opposite the variable x to build the resulting graph, as is done in Fig. 2.14. Then, we will see how the variable we choose changes over time.

After the checkbox is checked, the graph of interest will be automatically built in the selected axes (Fig. 2.31).

Now, we will build the phase diagram. To do this, instead of the time dependence, choose the dependence of one variable on another, as shown in Fig. 2.32.

Then, we get a graph as shown in Fig. 2.33.

Fig. 2.30 Selection of a graph showing time dependence

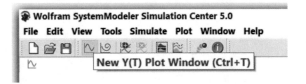

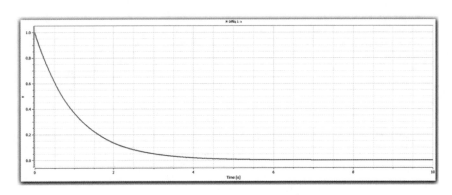

Fig. 2.31 Graph of *x*-coordinate versus time in a time interval of 10 s

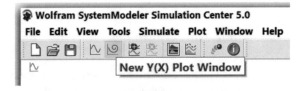

Fig. 2.32 Choice of dependence of an arbitrary variable *Y* on another variable *X*

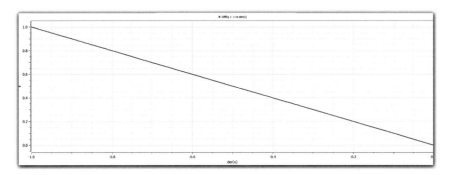

Fig. 2.33 Model phase trajectory

2.4.3 Creating a Library of Components for Complex Systems Modeling

It is very important to save the created model, since in the future we will most likely use it as a finished component in complex component systems. Return to the Model Center to create an icon for the model. Let's turn to the view mode as icons in the Class Window, by clicking on the Icon View button in the toolbar, as shown in Fig. 2.34.

To create an icon, use the drawing tools on the Model Center toolbar, as shown in Fig. 2.35.

To draw a rectangle, select the appropriate tool. Draw a rectangle covering the whole white area of the categories window. Double-click on the created rectangle to view and edit its properties, as shown in Fig. 2.36.

Change the fill color to Gray, select Solid the fill type, and click on the OK button. Next, press the Esc key to deselect. Finally, select the Text Tool, stretch the text input field to the entire rectangle, and then change the text to "Diff Eq" in the Text Properties dialog box. The reason for which you need to deselect before entering a text element requires explanation. Without deselecting, we would move the rectangle

Fig. 2.34 Icon View button on the Model Center toolbar

Fig. 2.35 Tools for drawing toolbar **Model Center**

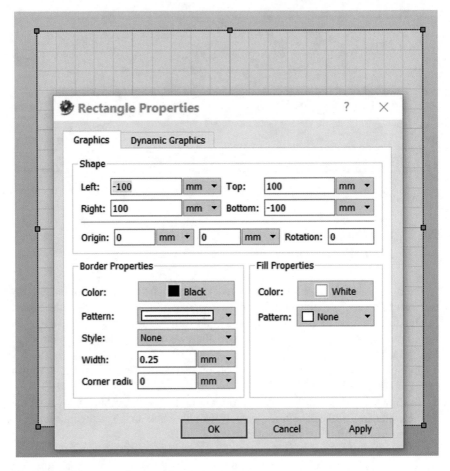

Fig. 2.36 Editing rectangle properties

instead of adding a text element, since all the drawing tools can also be used to move the selected elements.

Now, the icon of the new component should look like in Fig. 2.37.

Fig. 2.37 Icon of the created model

From now on, the Diff Eq component will be represented by this icon wherever it is used.

2.5 Basics of Component Modeling in WSM

In the third chapter, we will build mechanical models based on the differential equations describing this physical process, without resorting to standard libraries and the use of ready-made components. However, it is useful to understand how they can be used to significantly complicate the task.

In the previous section, we learned the basics of programming your own models or components in the Modelica language.

Let us now consider the process of creating a model using ready-made components from existing libraries. Let us find out how these two approaches are interconnected and how, having mastered the skills of writing code in the Modelica language, one can verify the correctness of the constructed component models.

Consider building a model that describes a simple one-dimensional spring system with damping. In the future, copying this model or its elements, you can create much more complex models of oscillatory systems. Perform modeling in two ways—by writing code in the Modelica language and using ready-made components from the built-in library.

2.5.1 Creating a Spring Pendulum Model

Let us write the equation of oscillations of the spring pendulum in the presence of medium resistance. This is a second-order differential equation:

$$m\frac{d^2x}{dt^2} + kx + \gamma\frac{dx}{dt} = 0$$

This equation can be solved analytically. A detailed theoretical analysis of this model will be given in the third chapter.

First, we write the code in the Modelica language without using ready-made components. Let's prepare this differential equation of the second order to solve using Wolfram SystemModeler package. To do this, we reduce it to a system of two first-order equations.

$$\frac{dx}{dt} = v$$

$$m\frac{dv}{dt} = -kx - \gamma v$$

We write the code on Modelica, providing the ability to change parameters such as spring stiffness and resistance coefficient and initial conditions—the offset from the equilibrium position and the initial velocity. To do this, we need to introduce into consideration concepts such as the type of data on the level of variability.

In the Modelica language, the principle of variability takes place; that is, all data has its own level of variability. By variability, data is divided into variables, parameters, and constants.

Variables. Used for time-dependent data, that is, their value changes during the execution of a computational experiment; a set of variables describes the state of the object being modeled. Variable declaration is as follows:

```
Real x;
```

Variables of the same type can be grouped together using the following syntax:

```
Real x, y;
```

A declaration may also be followed by a description, for example:

```
Real alpha "angular acceleration";
```

Parameters get their value before performing a computational experiment and do not change during it, often used to configure objects of the same class. Parameters are declared using the **parameter** keyword:

```
parameter Real x;
```

Constants. Constants, like parameters, are initialized before performing a computational experiment and do not change during it, but unlike parameters, their value cannot be "set" manually. Constants are declared using the **constant** keyword:

```
constant Real x;
```

When declaring variables, parameters and constants always indicate their type. Read more about the types of variables in [4].

So, to declare a parameter, we use **parameter**, to declare constants—constant. Variables can immediately set initial values using the **start** attribute after the variable name. There you can also set the dimension of the physical quantity. In our problem, the variables are the displacement and speed, with given initial conditions, and the parameters are body mass, spring stiffness, and medium viscosity (or damping coefficient). A constant for us could be, for example, the acceleration of free fall.

We introduce the parameters that can be changed in the future. We will indicate the dimension in brackets, after the equal sign—the initial value, at the end we will write the verbal description of the parameter.

```
Parameter Real m (unit = "kg") = 1 "mass";
Parameter Real k (unit = "N/m") = 100
 "coefficient of spring stiffness";
Parameter Real gamma (unit = "N s/m") = 1 "viscosity".
```

We introduce the variables. In this model, this is displacement and speed.

```
Real x (unit = "m", start = 0.05) "displacement";
Real v (unit = "m/s", start = 0.1) "velocity".
```

We write the defining equations. They are written in the **equation** block. The time derivative is specified using the **der()** function.

```
equation;
der(x) = v;
der(v) = -k / m * x - gamma * v.
```

As a result, we write the program code of the model, as shown in Fig. 2.38.

We have the opportunity to edit the system settings directly in the program code or by opening the parameters window at the bottom, as shown in Fig. 2.39.

It is possible to separately adjust the parameters and separately the initial values.

Before moving on to component modeling, we also mention the possibility of the Modelica language, such as creating functions. They are convenient for creating your

```
dampedspringpendulum

model dampedspringpendulum
   parameter Real m(unit = "kg") = 1 "mass";
   parameter Real k(unit = "N/m") = 100 "coefficient of spring stiffness";
   parameter Real gamma(unit = "N s/m") = 1 "viscosity";
   Real x(unit = "m", start = 0.05) "displacement";
   Real v(unit = "m/s", start = 0.1) "velocity";
equation
   der(x) = v;
   m * der(v) = -k * x - gamma * v;
   ¤;
end dampedspringpendulum;
```

Fig. 2.38 Program code in WSM written in Modelica

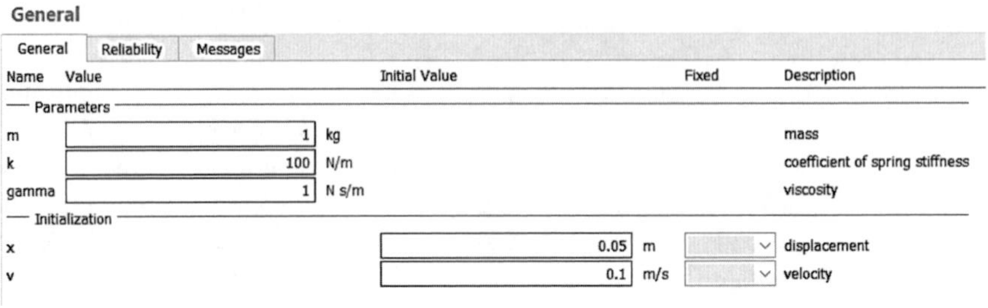

General

General	Reliability	Messages				
Name	Value		Initial Value		Fixed	Description
Parameters						
m	1	kg				mass
k	100	N/m				coefficient of spring stiffness
gamma	1	N s/m				viscosity
Initialization						
x			0.05	m	⌄	displacement
v			0.1	m/s	⌄	velocity

Fig. 2.39 Setting model parameters

own components, which use standard formulas for calculating physical, geometric, or other characteristics.

Addition: Functions

*The **function** class describes a function written in an algorithmic language. Function setting syntax:*

```
function Square
    // Declaring input and output variables
protected
    // Declaring intermediate variables (if required)
algorithm
    // Body function
end Square;
```

*Input variables are declared with the **input** attribute, and **output** variables are declared with the output attribute. For example, setting a function to calculate the volume of a cylinder:*

```
function CylinderVolume
    input Real radius;
    input Real length;
    output Real volume;
algorithm
    volume = 3.14159*radius^2*length;
end CylinderVolume;
```

You can call a function in several ways. You can explicitly specify the names of input variables when calling a function, or you can simply call a function with numeric values. In this case, the variables will get the values according to their order when they are declared. For example:

```
CylinderVolume (0.5,  12.0);   // radius=0.5, length=12.0
CylinderVolume (12.0,  0.5);   // radius=12.0, length=0.5
CylinderVolume                 // radius=12.0, length=0.5
(length=12.0,radius=0.5);
```

More complete information about the functions can be found in [4].

Let's turn to component modeling. We assemble the circuit from the finished components, setting the same initial values for the mass, stiffness, and initial values of the velocity and displacement.

Consider the rules for creating a diagram of the component model. To do this, we need to find suitable components, drag the found components into the diagram area, and connect all the components together using the Connection Line Tool.

That is, we need the availability of ready-made components and connecting interfaces that provide connection to other components. It will show how easy it is to add connecting interfaces using the Connection Line Tool. Also, certain parameters can be set for the components, which makes their further use more flexible.

The Modelica Connection Line Tool is a **connector** class, which is a special type of class that serves to communicate between components. A connector, unlike **model** or **block** types, cannot have equations and contains only variables involved in relationships. Common syntax for defining a connector:

```
     connector ConnectorName
"Description of the connector"
      // Variables are declared here.
     end ConnectorName;
```

If you plan to use a connector to create directional links, you must declare variables with input or output attributes to declare **input** and **output** variables, respectively.

For non-directional links, i.e., for physical modeling, there are two types of variables.

The first type is "potential" variables or contacts. Typical examples of potential variables are temperature, voltage, and pressure. Changes in these variables usually lead to dynamic behavior in the system. These variables implement the first Kirchhoff rule.

The second type is "pass-through" variables or streams. Flow variables are usually a flow of some conservative value, such as mass, momentum, energy, and charge. These flows are usually the result of some difference in all variables in the component model. For example, the current flowing through the resistor is through the voltage difference on the two sides of the resistor. These variables implement the second Kirchhoff rule.

Example of declaring a connector that can serve to connect hydraulic components:

```
connector Liquid
    Modelica.SIunits.Pressure p; // Pressure
    flow Modelica.SIunits.VolumeFlowRate q;
// Volume flow rate
  end Liquid;
```

More details about the description of connectors in the Modelica language can be found in [4].

So let's continue the creation of the component model. Create a new model, and open an empty layer in the **Diagram View**. Note that you need to go from Text to Diagram View.

To view the package structure of the built-in Modelica library, you need to use the **Class Browser** window.

Double-clicking on the package name will open its hierarchical structure and display the contents in the *Class Browser* window. Double-clicking on the model name will open the model in the Class Window. Additional information about models in the Modelica libraries is integrated into packages and models and can be displayed by right-clicking on any of the packages or models in the Class Browser window and then choosing *View Documentation* from the pop-up menu. This way you can get information about any component from the *Libraries Modelica* built-in library.

In addition to the built-in library *Libraries Modelica* and the library of *Examples*, the *Class Browser* window displays the latest models created by you and recently used components.

Open the **Modelica Standard Library,** and find the **Mechanics.Translation Library**, as shown in Fig. 2.40a.

Double-clicking on the *Class Documentation* button will open a window with full information on the model or package, as shown in Fig. 2.40b. On the other hand, Fig. 2.41 presents a window with documentation on the entire package of mechanical components of the built-in library *Libraries Modelica*.

Find the **Fixed** component in the **Components** section, and drag it onto the diagram layer, as shown in Fig. 2.42.

To make the component model diagram visually match the physical description of the model, you can rotate the **Fixed** component by right-clicking the component and selecting it, and then rotate it with the **Rotate Right** function (or simply press **Ctrl + R**), as shown in Fig. 2.43.

If everything was done correctly, then as a result you will see a component rotated 90°, i.e., modeling in this case the side mounting to the wall, as shown in Fig. 2.44.

The next step is to add components to simulate the spring and damper. You can use two different components; however, there is a component in the library that

(a)   **(b)**

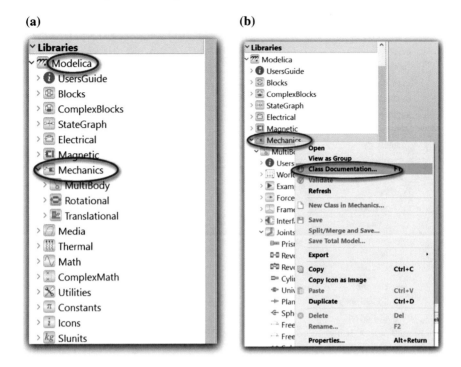

Fig. 2.40 Selecting the required component from the library

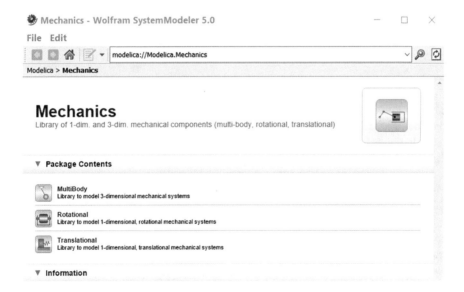

Fig. 2.41 Class documentation window

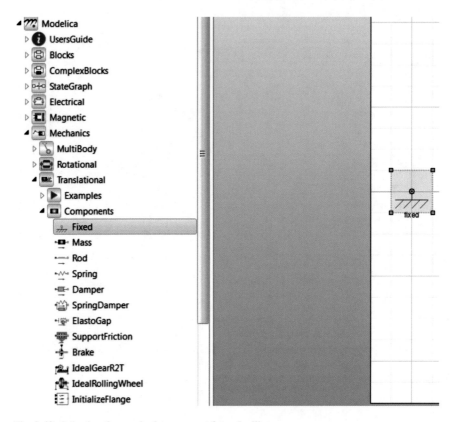

Fig. 2.42 Selecting the required component from the library

combines a spring and a damper—**SpringDamper**. Drag this component onto the chart layer. Now, we need to connect these two components together. This can be done by clicking the **Connection Line Tool** in the toolbar (or by pressing *C*), as shown in Fig. 2.45.

Only connectors that have similar properties can be interconnected. Support for this rule is implemented in the *Connection Line Tool*. If the user tries to connect two incompatible connectors, the connection will be blocked.

Multi-module connectors, called "frames," are a local coordinate system associated with a given component.

In our case, the spring with the damper is connected to the side mount and to the ground using one-dimensional mechanical system connectors, called "flanges." For example, to connect the spring **frame_a** flange with the mass **frame_b** flange, move the cursor to one flange, press the left mouse button, and, while holding the button down, move the cursor to another flange. To end the connection, release the mouse button.

The result should look like the illustration in Fig. 2.46.

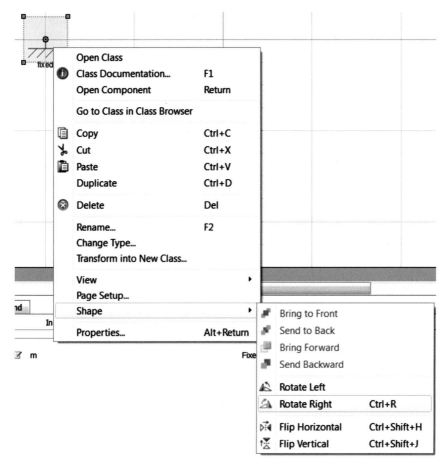

Fig. 2.43 Rotate the selected component

Fig. 2.44 Side attachment
component

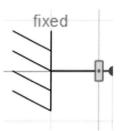

Fig. 2.45 Connecting
components with the
Connection Line Tool

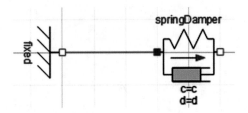

Fig. 2.46 Spring model with damper with fixed end

The next step is to add mass, which should simulate the body in the finished component model. Drag the **Mass** component onto the diagram layer, as shown in Fig. 2.47. Connect it with **SpringDamper**:

The default model is ready.

In the diagrammatic representation, this model has the structural view as shown in Fig. 2.48.

Configure the characteristics of the component. This can be done in the Text View. When we drag and connect components, the model editor generates Modelica code corresponding to the actions performed. Switch to Modelica Text View to see the textual representation of the model being created. In the textual representation of

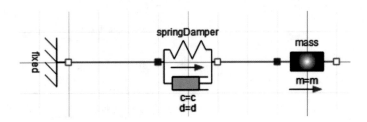

Fig. 2.47 Spring pendulum model

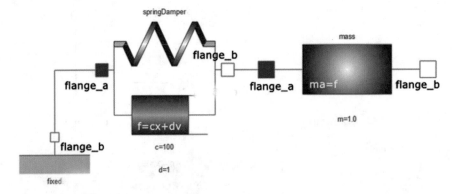

Fig. 2.48 Block diagram of the spring pendulum

the model, each of its components is declared, and each connection between the two components is represented by a linking equation, which are listed in the equation section.

However, it is more convenient to do this in graphical mode, switching to the mode of viewing component properties.

To do this, simply double-click on the component of interest. For mass, for example, we obtain the view as shown in Fig. 2.49.

View the code of this finished component by going to text mode, as shown in Fig. 2.50.

The equations describing body behavior look like the following:

$$v = \text{der}(s);$$
$$a = \text{der}(v);$$
$$m * a = \text{flange\_a.f} + \text{flange\_b.f.}$$

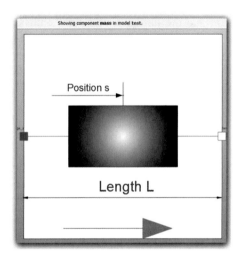

Fig. 2.49 View the mass component

```
Modelica.Mechanics.Translational.Components.Mass

model Mass "Sliding mass with inertia"
  parameter SI.Mass m(min = 0, start = 1) "Mass of the sliding mass";
  parameter StateSelect stateSelect = StateSelect.default "Priority to use s and v as states" ¤;
  extends Translational.Interfaces.PartialRigid(L = 0, s(start = 0, stateSelect = stateSelect));
  SI.Velocity v(start = 0, stateSelect = stateSelect) "Absolute velocity of component";
  SI.Acceleration a(start = 0) "Absolute acceleration of component";
equation
  v = der(s);
  a = der(v);
  m * a = flange_a.f + flange_b.f;
  ¤;
end Mass;
```

Fig. 2.50 View the program code for the mass component

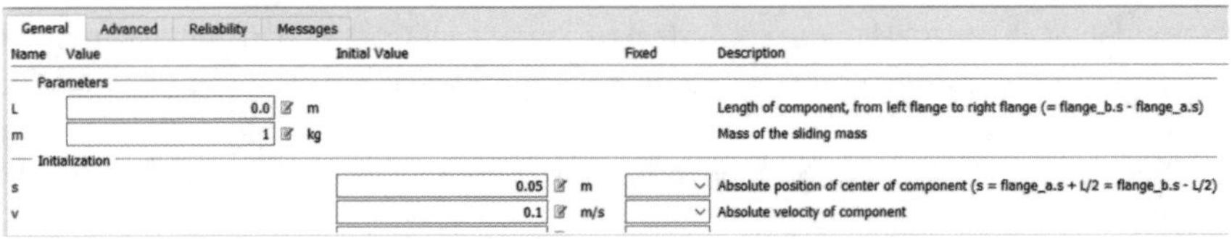

General	Advanced	Reliability	Messages			
Name	Value		Initial Value		Fixed	Description
Parameters						
L		0.0 ☑ m				Length of component, from left flange to right flange (= flange_b.s - flange_a.s)
m		1 ☑ kg				Mass of the sliding mass
Initialization						
s			0.05 ☑ m	∨		Absolute position of center of component (s = flange_a.s + L/2 = flange_b.s - L/2)
v			0.1 ☑ m/s	∨		Absolute velocity of component

Fig. 2.51 Entering numeric mass component parameters

The first two equations are differential, describing the relationship of displacement, velocity, and acceleration. They are universal and do not change depending on the type of process being modeled.

The third equation is Newton's second law in general. In different models, the mass can move due to different reasons and a specific equation here cannot be written.

To the right in this equation are the forces applied to the **flange_a** and **flange_b** connectors of this component. In our model, the mass is attached to the spring using the **flange_a** connector, and the **flange_b** connector remains free. Therefore, the force should be set as an output signal from the corresponding connector **flange_b** springs.

Before moving on to the spring, open the parameters window at the bottom. We introduce the necessary values. The geometrical dimensions of the mass will be assumed to be zero, since we consider mass to be a material point. Choose $m = 1$ kg. In addition, at the bottom of the window we set the initial conditions—the displacement of the center of mass by 0.05 m and its initial speed of 0.1 m/s, as shown in Fig. 2.51.

Next, configure the "SpringDamper" component—a spring with a damper, as shown in Fig. 2.52.

View the code of this finished component by going to the text mode, as shown in Fig. 2.53.

There are no equations of motion, but there is a definition of the elastic force of the spring and the resistance of the medium:

$$f\_c = c * (s\_rel - s\_rel0); \quad \text{elastic force}$$
$$f\_d = d * v\_rel; \quad \text{resistance force}$$
$$f = f\_c + f\_d; \quad \text{total force}$$

supplied from the **flange_b** connector to the **flange_a** connector

$$lossPower = f\_d * v\_rel; \quad \text{viscosity losses}$$

Substituting the force f from the "SpringDamper" component into the right-hand side of the equation m * a = flange_a.f + flange_b.f and adding differential equations v = der (s); a = der (v) from the "Mass" component, we arrive at the equations that we wrote down during direct programming in the Modelica language:

$$der(x) = v;$$

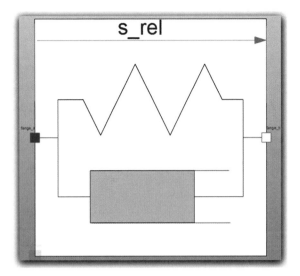

Fig. 2.52 View SpringDamper

```
Modelica.Mechanics.Translational.Components.SpringDamper
model SpringDamper "Linear 1D translational spring and damper in parallel"
   extends Translational.Interfaces.PartialCompliantWithRelativeStates;
   parameter SI.TranslationalSpringConstant c(final min = 0, start = 1) "Spring constant";
   parameter SI.TranslationalDampingConstant d(final min = 0, start = 1) "Damping constant";
   parameter SI.Position s_rel0 = 0 "Unstretched spring length";
   extends Modelica.Thermal.HeatTransfer.Interfaces.PartialElementaryConditionalHeatPortWithoutT;
protected
   Modelica.SIunits.Force f_c "Spring force";
   Modelica.SIunits.Force f_d "Damping force";
equation
   f_c = c * (s_rel - s_rel0);
   f_d = d * v_rel;
   f = f_c + f_d;
   lossPower = f_d * v_rel;
   □;
end SpringDamper;
```

Fig. 2.53 View the program code of the SpringDamper component

m * der(v) = (-k * x) - gamma * v.

Open the parameters window again. We introduce the necessary values. The spring stiffness is 100 N/m, and the length of the undeformed spring is assumed to be 10 cm. Given that we shifted the mass by 5 cm, the distance between the ends of the spring at the initial moment of time must be taken equal to 15 cm. That is, the spring is deformed at the initial moment of time. The coefficient of viscosity is set equal to 1 Ns/m (Fig. 2.54).

Configure the "Fixed" component, as shown in Fig. 2.55.

Using this component, we fix the left end of the spring at a fixed distance from the origin s0. In our example, we combined the origin of coordinates with the mass, so the mount is located at the point s0 = −0.10 m (Fig. 2.56).

General	Advanced	Reliability	Messages			
Name	Value		Initial Value		Fixed	Description
— Parameters						
useHeatPort	false ∨ ☑					=true, if heatPort is enabled
c	100 ☑ N/m					Spring constant
d	1 ☑ N·s/m					Damping constant
s_rel0	0.1 ☑ m					Unstretched spring length
— Initialization						
s_rel			0.15 ☑ m	∨		Relative distance (= flange_b.s - flange_a.s)

Fig. 2.54 Entering the numeric values of the parameters of the SpringDamper component

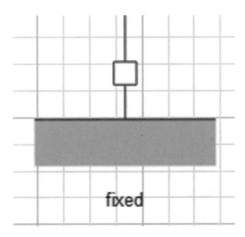

Fig. 2.55 View component "Fixed"

```
model Fixed "Fixed flange"
  parameter SI.Position s0 = 0 "Fixed offset position of housing";
  Interfaces.Flange_b flange ¤;
equation
  flange.s = s0;
  ¤;
end Fixed;
```

Fig. 2.56 Enter the numeric values of the parameters of the component "Fixed"

As a result, switching to the text mode, we see the finished program with explicitly defined connections between the components, as shown in Fig. 2.57.

Let's go into the mode of performing a numerical experiment. We construct in one axis the graphs of the displacement and velocity of the body in two models. We see almost complete overlay of the corresponding graphs, as shown in Fig. 2.58.

```
model test1
    Modelica.Mechanics.Translational.Components.Mass mass(L = 0.0, m = 1.0, s.start = 0.05, v.start = 0.1) ¤;
    Modelica.Mechanics.Translational.Components.Fixed fixed(s0 = -0.10) ¤;
    Modelica.Mechanics.Translational.Components.SpringDamper springDamper(c = 100, d = 1, s_rel0 = 0.1, s_rel.start = 0.15)
equation
    connect(fixed.flange, springDamper.flange_a) ¤;
    connect(springDamper.flange_b, mass.flange_a) ¤;
    ¤;
end test1;
```

Fig. 2.57 Spring pendulum simulation with damping using standard components

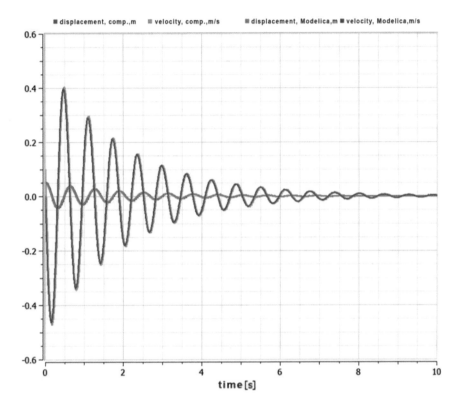

Fig. 2.58 Graphs of displacement and velocity versus time of a spring pendulum in a viscous medium using standard components and using program code

2.5.2 Analysis of the Properties of the Spring Pendulum Component

Let us analyze the above model of a spring pendulum by damping an important issue such as sensitivity analysis with respect to variable parameters.

When building a model, the parameters are most often selected manually with a certain degree of uncertainty, and further, the parameters may vary depending on external conditions and over time. Sensitivity means that with a small change in the input parameters, there is a change in the system properties that can be detected under conditions of computational error.

When analyzing, the sensitivity is determined by the changes in the response of the model to the deviations of individual parameters of the model. This allows us to conclude about the relative importance of input variables for a particular model, to identify key variables.

For example, when working with a spring pendulum, it would be useful to understand which parameter—it makes sense to change the mass or stiffness of the spring if you want to change the nature of the oscillations, setting a certain natural frequency.

This is done using the CVODES, which can perform an input sensitivity analysis. The sensitivity $s_i(t)$ for the state $y_i(t)$ with respect to the parameter p is determined by the expression:

$$s_i(t) = \frac{\partial y_i(t)}{\partial p}$$

In other words, at each time point, the sensitivity indicates how much the solution for the state $y_i(t)$ changes with the slightest change in the parameter p.

Consider the sensitivity of our oscillator with respect to two parameters of the system. To do this, select the CVODES in the experiment settings and check the SA cells on the Parameters tab for the mass and spring stiffness, as shown in Fig. 2.59.

Upon completion of the simulation, we can find the results of the sensitivity analysis on the tab of the plotter (Plot), inside the drop-down hierarchical tree of the system states, as shown in Fig. 2.60.

Perform a numerical sensitivity analysis experiment. The results are shown in Fig. 2.61.

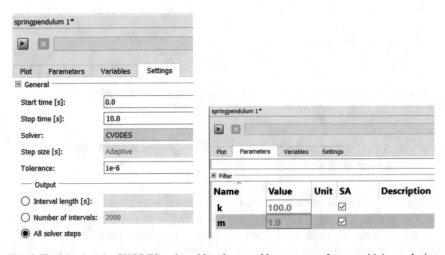

Fig. 2.59 Selecting the CVODES and marking the m and k parameters for a sensitivity analysis

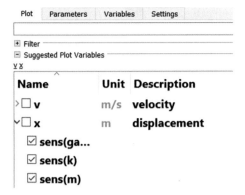

Fig. 2.60 Choice of sensitivity of the magnitude of the displacement of the cargo relative to the parameters of mass, spring stiffness, and viscosity

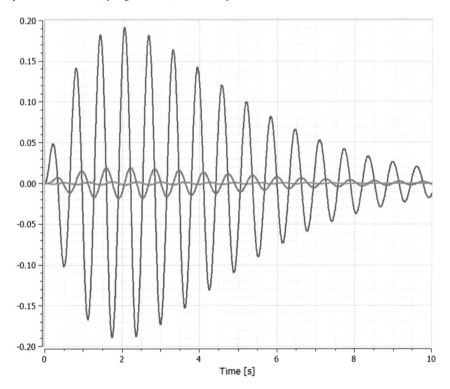

Fig. 2.61 Comparison of the effect of mass, spring stiffness, and medium viscosity on the system

Here, it can be seen that the spring stiffness has a negligible effect on the displacement during the entire simulation time. Mass has the most significant effect, and its effect increases when the time is equal 2 sec and decreases as the oscillations decay.

2.5.3 Example of Using the Spring Pendulum Component: Multi-link

As an example, we consider an articulated pendulum (this model can be easily built using the built-in library) and build on it a multi-link model. Please note that this example is also available in the **IntroductoryExamples** library. But we will create it anew, using the principles of component modeling.

The components needed to create a chain link are available in the *Modelica Standard Library*, which is part of **SystemModeler**. The basis of the chain link is a pendulum from the **Modelica.Mechanics.MultiBody** library, consisting of a body rotating around a pivot joint. To add friction to the rotation, a damper is connected to the pivot joint.

Start by creating a new model, which we will call ChainLink.

This will require components such as the Revolute hinge contained in the **Modelica.Mechanics.MultiBody** package in the *Joints* tab, the *BodyBox* body contained in the **Parts** tab, and the *Damper* damper located in the **Rotational.Components** tab, as shown in Fig. 2.62.

To add components to the **ChainLink** model, drag them from the **Class Browser** window and drop them onto the sheet in the **Diagram View**. On the sheet, they should look like those displayed in Fig. 2.63.

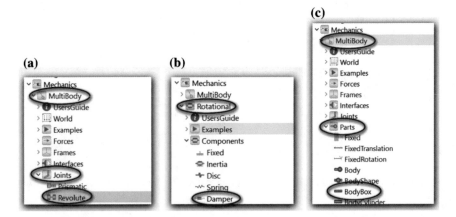

Fig. 2.62 An example of the selection of elements from the built-in library Modelica.Mechanics. **a** Revolute, **b** Damper, **c** BodyBox rod

(a)   **(b)**   **(c)**

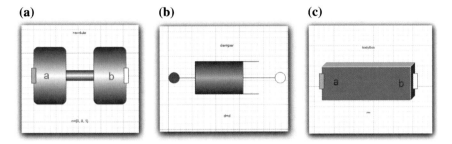

Fig. 2.63 Selected swivel components. **a** Revolute, **b** Damper, **c** BodyBox rod

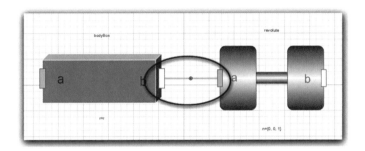

Fig. 2.64 An example of the operation of the Connection Line Tool

After adding components, they need to be interconnected. Components are connected using the Connection Line Tool ▯ on the Model Center toolbar, Fig. 2.64.

Since we want to use this model as a component for use in other more complex models, we need to add the appropriate connectors to connect the model with other components.

The resulting model is presented in Fig. 2.65.

As parameters, we will declare the dimension **r** of the **BodyBox** body **r** (length, width, height), as well as the damping coefficient **d** (Fig. 2.66).

Create an icon for this component. Switch to Icon View mode. and draw an icon using the Graphic Tools toolbar.

Note that connectors created using the Connection Line Tool were automatically added to the icon. Denote the chain link with an ellipse. To change the properties of the ellipse, double-click on the ellipse object, or select the ellipse with the mouse and press the Enter (Return) key. Use the Text Tool to specify the name of the component by adding a text element with the text "% name". To add a display of any available parameters, you must enter "%" followed by the name of the parameter, Fig. 2.67.

When the link component is at our disposal, the model of the articulated pendulum can be represented as a junction of four links interconnected and attached to the

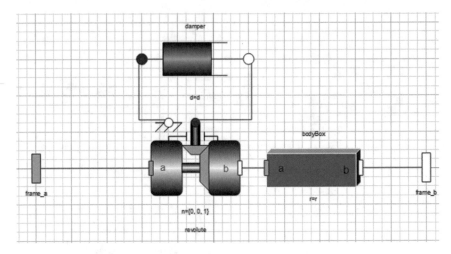

Fig. 2.65 Diagram View of ChainLink model

```
🗎                                              ChainLink                                              ⊠
ChainLink
model ChainLink
  Modelica.Mechanics.MultiBody.Parts.BodyBox bodyBox(r = r);
  Modelica.Mechanics.MultiBody.Joints.Revolute revolute(useAxisFlange = true);
  Modelica.Mechanics.Rotational.Components.Damper damper(d = d);
  Modelica.Mechanics.MultiBody.Interfaces.Frame_a frame_a;
  Modelica.Mechanics.MultiBody.Interfaces.Frame_b frame_b;
  parameter Real r[3] = {1, 0.1, 0.1};
  parameter Real d = 1.0;
equation
  connect(revolute.frame_b, bodyBox.frame_a);
  connect(frame_a, revolute.frame_a);
  connect(bodyBox.frame_b, frame_b);
  connect(damper.flange_a, revolute.support);
  connect(damper.flange_b, revolute.axis);
  ¤;
end ChainLink;
```

Fig. 2.66 Text View of the articulated pendulum model

Fig. 2.67 Chain link
component in Icon View

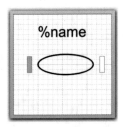

base. A diagrammatic representation of the articulated pendulum model is shown in Fig. 2.68.

We turn to the Simulation Center and simulate the model for 10 s. An animation of the pendulum's behavior can be displayed by clicking on the **Animation** button on the toolbar (Figs. 2.69 and 2.70).

Let's switch to graphing mode, as we did when creating our first simplest model of a differential equation. Let's display two dependencies on the screen at once—the x- and y-coordinates of the last link of the chain created by us from time, as shown in Fig. 2.71.

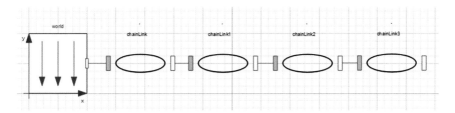

Fig. 2.68 Model of the articulated pendulum in the diagram representation

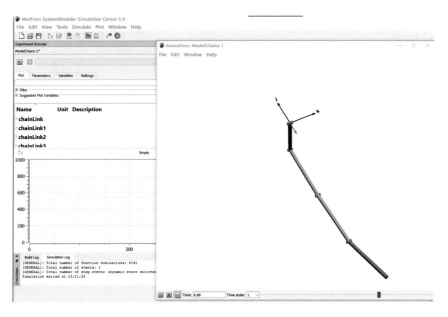

Fig. 2.69 General view of the screen when starting the animation model of the chain

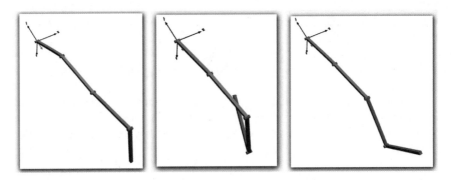

Fig. 2.70 Visualization of the articulated pendulum behavior at different points in time

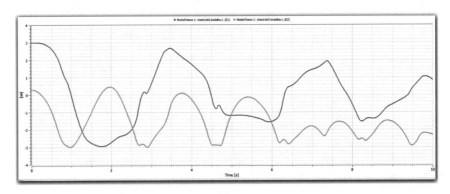

Fig. 2.71 Graph of horizontal and vertical position of the tip of the pendulum for 10 s

2.6 Creating a Hybrid Model in the WSM Package on the Modelica Language

In practice, one often has to deal with discrete–continuous models of dynamic systems, which are called hybrid systems [23, 24]. Other names of similar systems are "systems with a variable structure" and "event-driven systems." The variable structure of the models is due to the presence of slow (continuous) and fast (discrete) processes in the system. Thus, a distinctive feature of the complex behavior of the hybrid system is the set of qualitatively different and successively changing modes of operation. Mode or continuous behavior is called the state of the hybrid system, and mode switching is called discrete actions or events. The time taken to perform discrete actions is not taken into account; from the point of view of the functioning of the system, they are performed instantly.

Hybrid models allow systems to be described in a variety of engineering applications. For example, in mechanical objects, continuous motion can be interrupted or corrected by some physical impact. In addition, hybrid systems are used in the modeling of heterogeneous systems, the elements, and subsystems of which have a different physical nature.

In the classical representation, the continuous behavior of the HS is described by a system of differential or differential–algebraic equations. Discrete behavior is represented by logical predicates (finite automata) determining the states and conditions of transitions between continuous states.

In the Modelica language, states with continuous behavior (if there are several) can be described using the **if** statement, whereas for tracking events, the **when** statement is usually used. It should be noted that the event is described by a logical expression and is formed at that moment as soon as this expression changes the value from **false** to **true**. If, in this case, the trigger condition for the event continues to be fulfilled, it will not occur a second time until it returns from the trigger condition again.

2.6.1 Modelica Language Features to Describe Continuous and Discrete Events

To describe hybrid models with multiple states, use the **if** statement. The **if** operator allows you to switch between states that are described by different differential–algebraic equations, if the condition.

The **if** statement generally has the following syntax:

```
if cond1 then
    // This branch is used if cond1 == true
elseif cond2 then
    // This branch is used if cond1 == false
and cond2 == true
    // ...
elseif condn then
    // This branch is used if all previous
conditions are false
    // and condn == true
else
    // If all conditions are false,
    // then this branch is used
end if;
```

It is important to note that when the **if** operator is used, the number of equations must be the same regardless of which branch is being executed (this is also the case for **elseif**). One of the exceptions is the use of **if** within the original equation or the initial section of the algorithm, where the **else** condition is not required, since the number of equations should not be the same for both branches. Another notable exception is the use of **if** inside functions, where, again, there is no requirement that the number of equations be the same for both branches.

In the equations, you can also use the line form of the if statement.

```
if cexpr then expr1 else expr2
```

However, it can be in one of the parts of the equation. For example:

```
model Model1
   Real x;
   Boolean cond(start=true);
equation
   der(x) = if cond then 1 else -1;
end Model1;
```

For details on the **if** statement, see reference [4].

The **when** statement is necessary when it is necessary to perform instantaneous (with respect to model continuous time) actions. For example, it is necessary to make a jump in the value of a variable when a condition is met. There can be various equations in the actions of the when operator.

The **when** operator has the following general form:

```
when expr then
    // Actions
end when;
```

Often in the actions of the **when** statement, use the **reinit** function, which allows you to instantly change the value of a variable. The **reinit** function looks like this:

```
reinit(var, expr);
```

Here, **var** is a variable state whose value is to be changed and **expr** is an expression whose calculation will give a new value for the variable **var**. As an expression, there may be just a number.

It is worth mentioning the difference between **if** and **when** statements. Actions in the **when** statement become active only for a moment when the discrete switching condition becomes true. In other cases, the when statement has no effect on the behavior of the model. The **if** statement or expression remains active as long as the conditional expression is true. If a statement or expression containing **if** contains an **else** operator, then any branch will always be active.

For more information about the **when** operator, see [4].

2.6.2   An Example of Using Modelica Language Tools: Bouncing Ball

Let us now consider the simplest hybrid model using the example of a bouncing ball.

The model should describe the vertical movement of the ball, which falls due to gravity and bounces off the floor. The acceleration of gravity is constant and equal to $g = 9.8$ m/s^2. The height above the floor from which the ball begins to move is h. The speed of the ball is denoted as v.

The origin is placed on the floor. Coordinate of the ball above the floor is positive. The speed of the ball is positive when it moves up. The following system of two equations describes the free vertical movement of the ball:

$$\frac{dh}{dt} = v$$

$$\frac{dv}{dt} = -g$$

These are continuous equations.

The initial position and speed of the ball will be denoted, respectively, by h_0 and v_0. The ball bounces up when it touches the floor. It is assumed that in this case the direction of the speed of the ball is reversed and its value decreases by 20%; that is, when hitting the floor, the ball loses some of the stored energy. Suppose that before contact with the surface the speed of the ball was equal to v_b, and after the rebound from the surface it became v_a. The processes occurring at the time of the rebound can be neglected, that is, assume that the rebound occurs instantly. This means that a discrete action is performed at the time of the bounce.

$$v_a = 0.8v_b$$

For example, if the speed of the ball before the rebound is -10 m/s, then after the rebound it is 8 m/s. A negative speed sign indicates that the ball is falling, and a positive sign indicates that it is moving up after a rebound.

We already know how to write continuous equations in WSM. Let us consider in detail how to describe a discrete event. This can be done using the **when** statement and the override function **reinit()**, as shown below.

```
when h ≤ 0, then;
  reinit (v, -0.8 · v);
```

end when.

The **reinit** function has two arguments. The first argument is a continuous variable whose value we want to override, and the second is an expression of type **Real**. When the **reinit** function is called, the expression of the second argument is evaluated, and the resulting value is assigned to the variable that is the first argument.

The position of the ball during the onset of a discrete event will be very close to zero, but, due to computational errors, may be inaccurate. We must exclude the possibility of "failing" below the floor level. To do this, add another discrete action:

When $h \leq 0$, we set $h_a = 0$.

Now, the description of discrete actions in the program will look like this:

```
when h <= 0, then;
reinit (v, -0.8 · v);
reinit (h, 0);
end when.
```

So, this model has two continuous equations and one discrete event, and when triggered, two actions are performed.

Run WSM and build the model according to the rules described above. It will display like the one shown in Fig. 2.72.

Let's experiment in 20 s to make sure that the ball correctly stops bouncing and goes to rest on the floor, as shown in Fig. 2.73.

Let's see how the speed of the ball changes, Fig. 2.74.

We make one more small remark about discrete s. During the simulation, events are generated, for example, when the logical expression "if" is used by the equation, affecting its appearance. When an event occurs, the solver stops and repeats actions to find the exact point of the event in time. Such iterations (iterations) can take some time, and it is recommended to try to minimize them if it is possible.

BouncingBall

```
model BouncingBall
Modelica.SIunits.Distance x(start=10, fixed=true);
Modelica.SIunits.Velocity v(start=0, fixed=true);
parameter Real c = 0.8 "Coeff. elastic bouncing";
parameter Modelica.SIunits.Acceleration g = 9.8;
equation
der(v) = -g;
der(x) = v;
when x <= 0 then
reinit(v, -c*v);
reinit(x, 0);
end when;
  ¤;
end BouncingBall;
```

Fig. 2.72 Bouncing ball model

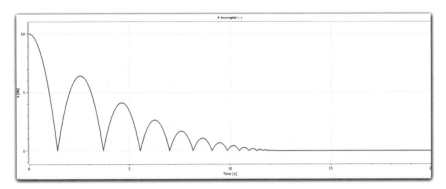

Fig. 2.73 Graph of the position of the bouncing ball above the floor

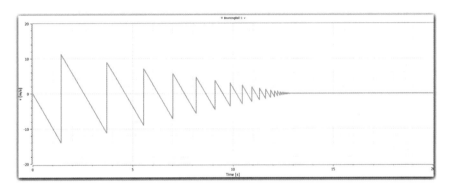

Fig. 2.74 Bouncing ball speed change

One way to avoid unnecessary iterations is to use the built-in **noEvent** operator, indicating to the solver that the generation of an event is not required. This operator can be used only if the expression is continuous during the occurrence of the event (it does not have to be differentiable).

```
model EventTest;
Real x1;
Real x2;
equation;
x1=if time < 3 then time else 3 "Event generated at time=3";
x2=noEvent (if time < 3 then time else 3) "No event generated";
end EventTest.
```

The **noEvent** operator can also be used to protect against incorrect calculations. The following example calculates **sin(x)/x**. To eliminate division by zero, **noEvent** is used to test the condition **abs(x)>0** to make sure that **sin(x)/x** is not calculated at **x=0**

```
Model GuardEval;
Real x;
```

```
Real y;
equation;
x=time - 1;
y=if noEvent(abs(x) > 0) then sin(x)/x else 1;
end GuardEval.
```

2.7 General Recommendations for Creating a Computer Model

We formulate a few general recommendations for modeling in the WSM environment. They may come in handy to avoid some common problems.

- Debugging code based on equations, compared to debugging code based on algorithms, is very different, since it is impossible to add breakpoints, perform a step-by-step solution, etc. Therefore, try to create your model step by step, while conducting small test components of the whole model. If an error occurs, it will be much easier to find it if the model is small.
- When creating components, use the Validate Class function described above to test them. If the component check was successful, then it is likely that it will work correctly with other components. A full-fledged component during verification should always return the same number of variables and equations. Note that an incomplete model often returns a different number of equations and variables.
- If the simulation takes a long time, this may be due to a large number of events. The number of events can be found in the Simulation Log after it is completed. If possible, try reducing the number of events using the noEvent operator or try changing your model. Note that when simulating discrete systems, an event is generated for each sample interval, which is normal.
- Try to avoid automatic adjustment of output intervals when modeling a familiar system. Since the data is written to the file for each step of the solver, this can significantly reduce the performance of the simulation of large systems. If you are familiar with the model's behavior, it is recommended to switch to a fixed number of intervals for the output signal or use the interval duration setting instead.

References

1. Wolfram SystemModeler: https://www.wolfram.com/system-modeler/
2. Modelica: https://modelica.org/
3. OpenModelica: https://openmodelica.org/
4. M.V. Tiller, Modelica by Example: https://mbe.modelica.university/
5. S.A. Agafonov, T.V. Muratova, Obyknovennye differencialnye uravneniya: ucheb. posobie dlya stud. vuzov. – Izdatelskij centr «Akademiya», Moscow (2008)

6. V.V. Amelkin, Differencialnye uravneniya v prilozheniyah. – Nauka, Gl. red. fiz-mat. lit., Moscow (1987)
7. A.I. Egorov, Obyknovennye differencialnye uravneniya s prilozheniyami, 2-e izd. – Fizmatlit, Moscow (2005)
8. V.V. Nemyckij, V.V. Stepanov, Kachestvennaya teoriya differencialnyh uravnenij, 2-e izd., pererab. i dop. – Gosudarstvennoe izdatelstvo texniko-teoreticheskoj literatury, Moscow-Leningrad (1949)
9. I.G. Petrovsky, Lectures On Partial Differential Equation. – Dover Publications Inc. (1992)
10. L.S. Pontryagin, Znakomstvo s vysshej matematikoj: Differencialnye uravneniya i ix prilozheniya. – Gl. red. fiz-mat. lit., Moscow (1988)
11. A.A. Samarskii, A.V. Gulin, Chislennye metody, Uchebnoe posobie. – Nauka, Moscow (1989)
12. E. Haier, S.P. Norsett, G. Wanner, Solving Ordinary Differential Equations I. Nonstiff Problems, 2nd edition, Springer Series in Computational Mathematics. – Springer-Verlag Berlin Heidelberg (1993)
13. E. Haier, G. Wanner, Solving Ordinary Differential Equations II. Stiff and Differential-Algebraic Problems, 2nd edition, Springer Series in Computational Mathematics. – Springer-Verlag Berlin Heidelberg (1996)
14. G. Hall, James Murray Watt, Modern Numerical Methods for Ordinary Differential Equations. – Clarendon Press (1976)
15. L.P. Shilnikov, A.L. Shilnikov, D.V. Turaev, L. Chua, Metody kachestvennoj teorii v nelinejnoj dinamike. Chast 1, Chast 2. –Institut kompyuternyx issledovanij, Moskva-Izhevsk (2004)
16. D. K. Arrowsmith, C. M. Place, Ordinary differential equations: a qualitative approach with applications. – Chapman and Hall, London and New York (1982)
17. A.C. Hindmarsh, L.R. Petzold, Algorithms and software for ordinary differential equations and differential-algebraic equations, Part 1: Euler methods and error estimation. – Comput. Phys. **9**(1), 34–41 (1995)
18. A.C. Hindmarsh, L.R. Petzold, Algorithms and software for ordinary differential equations and differential-algebraic equations, Part II: higher-order methods and software packages. – Comput. Phys. **9**(2), 48–155 (1995)
19. Numerical Methods for Solving Differential Equations. Heun's Method (Mathematics & Science Learning Center, Computer Laboratory): https://calculslab.deltacollege.edu/ODE/7-C-2/7-C-2-h
20. L.R. Petzold, A description of DASSL: a differential/algebraic system solver, in 10th International Mathematics and Computers Simulation Congress on Systems Simulation and Scientific Computation, Montreal, Canada, August 9, 1982, Sandia Report, 1982, ed. by L.R. Petzold, 10 p
21. R. Serban, A.C. Hindmash, CVODES, The sensitivity-enabled ODE solver in SUNDIALS, in ASME 2005 International Design Engineering Technical Conferences and Computers and Information in Engineering Conference, vol. 6: 5th International Conference on Multibody Systems, Nonlinear Dynamics, and Control, Parts A, B, and C, Long Beach, CA, USA, September 24–28, 2005 (ASME, 2005), pp. 257–269
22. R. Serban, A.C. Hindmash, CVODES: An ODE Solver with Sensitivity Analysis Capabilities. University of California, Lawrence Livermore National Laboratory, Technical Information Department's Digital Library, Livermore, CA (Preprint UCRL-JP-200039)
23. Yu.B. Kolesov, Yu.B. Senichenkov, Matematicheskoe modelirovanie gibridnyh dinamicheskih system. – Izdatelstvo Politehniceskogo universiteta, St. Petersburg (2014)
24. Yu.B. Kolesov, Yu.B. Senichenkov, Modelirovanie sistem. Dinamicheskie i gibridnye sistemy. – BHV-Peterburg, St. Petersburg (2012)

Chapter 3
Computer Simulation of Dynamic Systems

3.1 Dynamic System Modeling

A dynamic system is understood to mean any system, generally speaking, not only mechanical, the state of which can change over time [1–7]. A dynamic system is one of the most common mathematical models. This model arises when it becomes necessary to describe the mechanical motion of bodies in space under the action of applied forces. The position of the body is determined by its spatial coordinates, which continuously change from the initial value to the final value in the observation interval. The equations of the mathematical model describing the movement of the body give rise to the trajectory of its movement.

One of the simplest examples of mechanical dynamical systems is a material point capable of moving in three-dimensional space. The point state is set by six quantities. Three of these quantities are the Cartesian coordinates x, y, z; the other three are the corresponding components of the point velocity vector.

The time-dependent quantities that define the state of a dynamic system are called state variables or dynamic variables. In a theoretical study of any dynamic system, an operator is used, with the help of which they seek to find out how its state changes over time. The form of the operator may be differential, integral, or some other. Further, we will focus only on operators represented in differential form, that is, dynamic systems will be considered, the behavior of which is determined by the solutions of differential equations. The choice of the analytical description method sets, as is customary to say, the mathematical model of a dynamic system.

A model of a dynamic system describes an object or process for which the concept of a state is uniquely defined as a set of variables characterizing it at a given moment in time, and a law is given that describes the change (evolution) of the initial state over time. This law allows for the initial state to clearly predict the future state of a dynamic system. This law is often called the law of system evolution.

© Springer Nature Singapore Pte Ltd. 2020
K. Rozhdestvensky et al., *Computer Modeling and Simulation of Dynamic Systems Using Wolfram SystemModeler*,
https://doi.org/10.1007/978-981-15-2803-3_3

The main property of a dynamic system is that, knowing its state at a certain moment in time, it is possible to find a state at any subsequent moment in time. For this, it is enough to apply the law of evolution to the initial state.

The mathematical model of a dynamic system is considered given if the parameters (coordinates) of the system are entered that determine its state unambiguously and the law of state evolution in time is indicated.

Depending on the degree to which various factors are considered, different mathematical models can be associated with the same dynamic system.

Denote the coordinates of the system by $\mathbf{x} = (x_1, x_2, \ldots, x_n)$. These coordinates are called state variables or phase variables. The time coordinate is denoted by t. The law of evolution of a dynamic system in time in general is a system of algebraic–differential equations

$$F(\mathbf{x}, \mu, t) = 0$$

In this equation, the only independent variable is time t, and the state variables x depend on time, i.e., $\mathbf{x} = \mathbf{x}(t)$, μ is the vector of parameters that are assumed to be independent of time t, $F = (F_1, F_2, \ldots, F_n)$ are some given functions that, generally speaking, can be nonlinear and include time derivatives of variables state x.

Naturally, other types of equations can also be used, for example, partial differential equations. Then, some of the state variables can become independent variables.

If we consider the quantities $x_1, x_2, \ldots, x_n$ as the coordinates of the point $\mathbf{x}$ in the n-dimensional space $\mathfrak{R}^n$, we get a clear geometric representation of the state of the dynamical system in the form of this point. This point is called the image or phase point, and the state space is called the phase space of the system. The change in the state of the system in time corresponds to the movement of the phase point along a line called the phase trajectory.

In this tutorial, we restrict ourselves to considering dynamical systems that can be specified by a mathematical model described by a system of ordinary differential equations of the first order:

$$\frac{d\mathbf{x}}{dt} = F(\mathbf{x}, \mu, t)$$

where t is time, $\mathbf{x} = \mathbf{x}(t)$ are state variables (phase variables), μ is a vector of time-independent parameters, $F = (F, F_2, \ldots, F_n)$ are some given functions that can be interpreted as phase velocity points in the space $\mathfrak{R}^n$.

As a rule, even complex engineering problems allow the use of mathematical models of this kind.

3.2 Fundamental Principles for Mathematical Models Development

3.2.1 Newton's Laws and Conservation Laws

Undoubtedly, one of the easiest ways to obtain mathematical models of mechanical systems is the use of fundamental laws of nature, such as Newton's second law and conservation laws (of matter, energy, momentum) [8].

Let us examine the application of these laws by example.

An ideal spring pendulum (an ideal oscillator) is a mechanical system consisting of a spring with a given coefficient of elasticity (stiffness) k, one end of which is rigidly fixed, and the second is attached to a load in the form of a material point with a given mass m, making harmonic oscillations.

When the spring is not deformed, the body is in equilibrium. If the body is removed from the equilibrium position (to stretch or compress the spring), the elastic force from the side of the deformed spring will act on it, returning the body to the equilibrium position, as shown in Fig. 3.1. The result is harmonic undamped oscillations.

Based on Newton's second law and Hooke's law, the behavior of an ideal spring pendulum is described by the equation:

$$m \frac{d^2 x}{dt^2} + kx = 0$$

with given initial conditions:

$$x(t = 0) = x_0,$$

$$\upsilon(t = 0) = \upsilon_0.$$

For the recorded problem, an analytical solution of the following general form can easily be obtained:

$$x = A \sin(\omega t) + B \cos(\omega t),$$

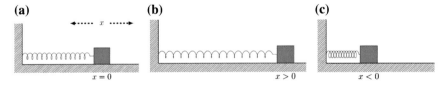

(a)   **(b)**   **(c)**

Fig. 3.1 Three positions of the spring pendulum: **a** neutral, **b** compressed, and **c** extended

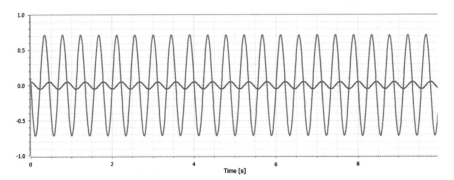

Fig. 3.2 Graphs of the bias and velocity versus time of an ideal harmonic oscillator ($m = 0.5$ kg, $k = 100$ N/m)

where the values of the constants A and B are determined taking into account the initial conditions (based on the initial state of the object). It is not difficult to show that the natural frequency of oscillations of the spring–mass system is $\omega = \sqrt{\frac{k}{m}}$, and the oscillation period of the spring pendulum is determined by the formula $T = 2\pi \sqrt{\frac{m}{k}}$. Graphs of the bias and velocity versus time of an ideal harmonic oscillator are sinusoids and are shown in Fig. 3.2.

The laws of dynamics with which this model was built should not contradict other fundamental laws of nature. If it is possible to verify this fact within the framework of the model you created, then it always makes sense to perform such a check.

In this case, to derive the dynamic law of motion of the load on the spring, you can use not Newton's law, but the energy conservation law. Since the attachment point of the spring is fixed, the wall does not work on the system and vice versa, the load on the spring also does not work. Therefore, the total mechanical energy E of the system remains constant. We calculate it.

Kinetic energy is determined by the movement of the ball (the spring is considered weightless):

$$E_k = \frac{m v^2}{2} = \frac{m}{2} \left(\frac{dr}{dt} \right)^2$$

It is easy to find the potential energy of the system by determining the work required to stretch or compress the spring by r:

$$E_p = -\int_0^r F \, dr' = -\int_0^r (-kr') \, dr' = \int_0^r kr' \, dr' = k \frac{r^2}{2}$$

For the total energy of the system that does not change with time $E = E_k + E_p$ (energy integral), we obtain

$$E = \frac{m}{2}\left(\frac{dr}{dt}\right)^2 + k\frac{r^2}{2}$$

Since $dE/dt \equiv 0$, differentiating the energy integral with respect to t, we arrive at the expression

$$m\frac{dr}{dt}\frac{d^2r}{dt^2} + k\frac{dr}{dt}r = \frac{dr}{dt}\left(m\frac{d^2r}{dt^2} + kr\right) = 0$$

So, we get the same equation as from Newton's law.

Hereinafter, all considered models are implemented in the Wolfram SystemModeler environment in Modelica language. Examples of the use of component modeling are discussed separately. To write code in Modelica, we write one second-order equation in the form of a system of two first-order equations:

$$\frac{dx}{dt} = v$$

$$m\frac{dv}{dt} = -kx$$

In this case, the program code looks like in Fig. 3.3.

Run a numerical experiment in the Simulation Center. Open the Experiment Browser simultaneously with the charting window. The screen view is shown in Fig. 3.4. Make sure from the analysis of the graphs that the total energy of the system really remains constant (red line).

```
springpendulum
model springpendulum
    parameter Real m ( unit = "kg" ) = 1 "mass";
    parameter Real k ( unit = "N/m" )= 100 "coefficient of spring stiffness";
    Real x(unit = "m", start = 0.05) "displacement";
    Real v(unit = "m/s", start = 0.1) "velocity";
    Real Ep(unit="J") "potential energy";
    Real Ek(unit="J") "kinetic energy";
    Real E(unit="J") "full energy";
equation
    der(x) = v;
    der(v) = -k / m * x;
    Ep = k*x*x/2;
    Ek = m*v*v/2;
    E = Ep + Ek;
    □;
end springpendulum;
```

Fig. 3.3 Program code that allows, in addition to studying displacement and velocity, calculating the potential, kinetic, and total energy of a harmonic oscillator

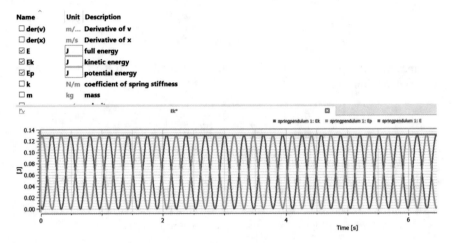

Name	Unit	Description
☐ der(v)	m/...	Derivative of v
☐ der(x)	m/s	Derivative of x
☑ E	J	full energy
☑ Ek	J	kinetic energy
☑ Ep	J	potential energy
☐ k	N/m	coefficient of spring stiffness
☐ m	kg	mass

Fig. 3.4   Graph of potential, kinetic, and total energy of a harmonic oscillator

In some cases, it is useful to consider the motion of the system in phase space. Phase space is the space formed by the variables that characterize the motion of a dynamic system.

Let us explain the concept of phase space by the example of a harmonic oscillator.

For the system under consideration, the phase space is two-dimensional and has two independent coordinates x and v. At each moment of time, the coordinate and velocity of the particle have a certain value and determine a certain point in the phase space, which is called the phase point. The phase point uniquely determines the state of the system at a given point in time. If we sequentially depict the phase points of the system for different instants of time, then we obtain a certain curve called the phase trajectory. In the case of a harmonic oscillator, the phase trajectory is an ellipse. Indeed, let the energy E of the oscillatory system remains constant, then separating both sides of the equation

$$E = \frac{m}{2}\left(\frac{dr}{dt}\right)^2 + k\frac{r^2}{2}$$

to constant E, we bring it to the form

$$\frac{x^2}{2E/k} + \frac{v^2}{2E/m} = 1$$

This is an ellipse equation with semi-axes $a = \sqrt{2E/k}$ and $b = \sqrt{2E/m}$ (see Fig. 1.2), which are determined by the stored energy and, therefore, given initial conditions. During oscillations, the coordinate $x = x_0\omega_0 \sin \omega_0 t$ changes, so that the phase point, moving clockwise, as shown in Fig. 3.5, performs a complete revolution in one oscillation period $T = 2\pi/\omega_0$.

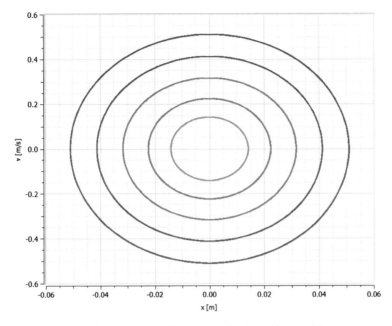

Fig. 3.5 Phase portrait of a harmonic oscillator ($\omega_0 = 2$ rad/s) with a special center-type point

If the system makes a periodic movement, a dot on the phase plane describes a closed curve, moving along it in a clockwise direction. The phase trajectory of the periodic motion is closed, because the system returns to its original mechanical state after each cycle of oscillations.

Generally speaking, only one phase trajectory passes through each point of the phase plane: If this point is chosen as the initial state of the system, the further movement of the system will be determined uniquely in accordance with the uniqueness of the solution of the Cauchy problem for the differential equation of the system. This movement will occur along the phase trajectory passing through a given point in the phase plane. In other words, the phase trajectories of the system do not intersect. The exception is only individual, isolated points of the phase plane. Such points through which more than one phase trajectory passes or no path passes are called special.

Thus, the phase portrait of a harmonic oscillator is a set of concentric ellipses, the size of which increases as a smooth function of energy. The phase portrait of a harmonic oscillator contains a singular point of the type "center." It is characteristic of undamped oscillations near the equilibrium position.

3.2.2 Variational Principles

Along with the fundamental laws for constructing models, variational principles are used that are comparable with them in terms of breadth of application and universality [8]. Variational principles are based on the consideration of rather general statements about the object under study, which state that only those that satisfy a certain condition are selected from all possible variants of its behavior. Usually, the condition indicates that the value associated with the object should reach an extreme value when passing from one state to another. These include, for example, the principle of possible movements and the principle of least action.

Let us explain the application of variational principles by the example of the Hamilton principle. To do this, make a brief description of it.

Let there be a mechanical system, the formal and strict definition of which we will not give yet, bearing in mind, however, that all interactions between the elements of such a system are determined by the laws of mechanics. We introduce the concept of generalized coordinates $Q(t)$ that completely determine the position of a mechanical system in space. The quantity $Q(t)$ can be a Cartesian coordinate (e.g., the r coordinate in the "load on spring" system), radius vector, angular coordinate, a set of coordinates of material points that make up the system, etc. It is natural to call dQ/dt the generalized velocity of a mechanical system at time t. The set of quantities $Q(t)$ and dQ/dt determines the state of the mechanical system at all instants of time.

To describe the mechanical system, the Lagrange function is introduced. In the simplest cases, the Lagrange function has a clear meaning and is written as

$$L(Q, dQ/dt) = E_k - E_p,$$

where E_k, E_p—kinetic and potential energies of the system, respectively. For the purposes of this problem, there is no need to give a general definition of the quantities E_k, E_p, since in the examples considered they are calculated in an obvious way.

We introduce the quantity $S[Q]$, called the action:

$$S[Q] = \int_{t_1}^{t_2} L\left(Q, \frac{dQ}{dt}\right) dt$$

The last integral, obviously, is a functional of the generalized coordinate $Q(t)$, i.e., of the function $Q(t)$ defined on the interval $[t_1, t_2]$, he associates a certain number S (action).

The Hamilton principle for a mechanical system states: If the system moves according to the laws of mechanics, then $Q(t)$ is a stationary function for $S[Q]$, or

$$\frac{d}{d\varepsilon} S[Q + \varepsilon\varphi] = 0, \quad \varepsilon = 0$$

The function $\varphi(t)$ that appears in the principle of least action is a certain test function that vanishes at time instants t_1, t_2 and satisfies the condition that $Q(t) + \varepsilon\varphi(t)$ is the possible coordinate of this system (otherwise, $\varphi(t)$ arbitrary).

The meaning of the principle of least action is that of all a priori admissible trajectories (movements) of the system between the moments t_1, t_2, a motion is selected (implemented) that delivers a minimum to the action functional (the name of the principle also comes from this). The function $\varepsilon\varphi(t)$ is called a variation of the quantity $Q(t)$.

So, the application scheme of the Hamilton principle for constructing models of mechanical systems is as follows: Generalized coordinates $Q(t)$ and generalized velocities dQ/dt of the system are determined, the Lagrange function $L(Q, dQ/dt)$ and the action functional $S[Q]$ are constructed, minimization which, on the variations $\varepsilon\varphi(t)$ of the coordinate $Q(t)$, gives the desired model.

Now consider the application of the Hamilton principle on the example of the "load on spring" system from the previous paragraph. As a generalized coordinate, it is convenient to choose the usual Cartesian coordinate of the load $r(t)$. Then, the generalized velocity $dr/dt = v(t)$ is the usual speed of the load. The Lagrange function, equal to $L = E_k - E_p$, is written in terms of the kinetic and potential energy of the system already found in the previous paragraph:

$$L = \frac{m}{2}\left(\frac{dr}{dt}\right)^2 - k\frac{r^2}{2}$$

For the magnitude of the action $S[r]$, we obtain the following expression:

$$S[r] = \int_{t_1}^{t_2} L\left(r, \frac{dr}{dt}\right) dt = \int_{t_1}^{t_2}\left[\frac{m}{2}\left(\frac{dr}{dt}\right)^2 - \frac{k}{2}r^2\right] dt$$

Now we calculate the action on the variations $\varepsilon\varphi(t)$ of the coordinate $r(t)$:

$$S[r + \varepsilon\varphi] = \int_{t_1}^{t_2} L\left(r, \frac{dr}{dt}\right) dt = \int_{t_1}^{t_2}\left[\frac{m}{2}\left(\frac{d(r + \varepsilon\varphi)}{dt}\right)^2 - \frac{k}{2}(r + \varepsilon\varphi)^2\right] dt$$

Now, differentiating the resulting formula with respect to ε and setting $\varepsilon = 0$, we obtain:

$$\left.\frac{d}{d\varepsilon}S[r + \varepsilon\varphi]\right|_{\varepsilon=0} = \int_{t_1}^{t_2}\left[m\frac{dr}{dt}\frac{d\varphi}{dt} - kr\varphi\right] dt$$

We integrate this expression in parts, taking into account the fact that $\varphi = 0$ at moments t_1 and t_2 and equate it to zero in accordance with the Hamilton principle:

$$\frac{d}{d\varepsilon} S[r + \varepsilon\varphi]\Big|_{\varepsilon=0} = -\int_{t_1}^{t_2} \varphi\left[m\frac{d^2 r}{dt^2} + kr\right]dt = 0$$

Since the test function $\varphi(t)$ is arbitrary, the part of the expression under the integral sign in square brackets must be equal to zero at all time $t_1 < t < t_2$:

$$m\frac{d^2 r}{dt^2} + kr = 0$$

such that the motion of the system must be described by the same equation that we obtained in the previous paragraph from Newton's law and the law of conservation of energy. All three approaches are equivalent.

Consider another example of the application of the Hamilton principle.

Consider a mathematical pendulum—that is, a harmonic oscillator—which is a mechanical system consisting of a material point at the end of a weightless inextensible thread or a weightless inextensible rod and located in a uniform field of gravity. The pendulum moves in a medium without resistance.

The returning force when the material point deviates from the equilibrium position is the projection of gravity onto the tangent to the trajectory along which this material point moves. The restoring force acting on the material point causes it to oscillate around the equilibrium position, as shown in Fig. 3.6.

After all simplifying assumptions, it is clear that the position of the pendulum is determined by only one generalized coordinate, for which we choose the angle $\theta(t)$ of the deviation of the rod from the vertical. The generalized velocity in this case is the angular velocity $d\theta/dt$.

The kinetic energy of the system is given by:

Fig. 3.6 Forces acting on a mathematical pendulum

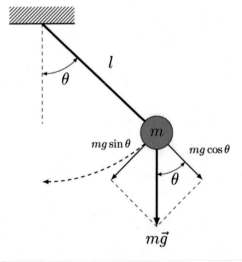

$$E_k = \frac{1}{2}mv^2 = \frac{1}{2}m\left(l\frac{d\theta}{dt}\right)^2 = \frac{1}{2}ml^2\left(\frac{d\theta}{dt}\right)^2,$$

and potential energy by the expression:

$$E_p = mgh = -mg(l\cos\theta - l),$$

where h is the deviation of the pendulum from the lowest vertical position. In further calculations, we omit the value of mgh in E_p, since the potential energy is determined accurate to a constant.

Now it is easy to calculate the Lagrange function

$$L\left(Q, \frac{dQ}{dt}\right) = E_k - E_p$$

and action

$$S[Q] = \int_{t_1}^{t_2} L\left(Q, \frac{dQ}{dt}\right) dt$$

with selected generalized coordinate and speed:

$$L\left(\theta, \frac{d\theta}{dt}\right) = ml\left[\frac{1}{2}l\left(\frac{d\theta}{dt}\right)^2 + g\cos\theta\right]$$

$$S[\theta] = ml\int_{t_1}^{t_2}\left[\frac{1}{2}l\left(\frac{d\theta}{dt}\right)^2 + g\cos\theta\right]dt$$

Finding action on variations $\alpha + \varepsilon\varphi(t)$:

$$S[\theta + \varepsilon\varphi] = ml\int_{t_1}^{t_2}\left[\frac{1}{2}l\left(\frac{d\theta}{dt} + \varepsilon\frac{d\varphi}{dt}\right)^2 + g\cos(\theta + \varepsilon\varphi)\right]dt$$

$$= ml\int_{t_1}^{t_2}\left[\frac{1}{2}\{\left(\frac{d\theta}{dt}\right)^2 + 2\varepsilon\frac{d\theta}{dt}\frac{d\varphi}{dt} + \varepsilon^2\left(\frac{d\varphi}{dt}\right)^2 + g(\cos\theta + \varepsilon\varphi)\right]dt$$

Differentiating with respect to ε and setting $\varepsilon = 0$, we obtain

$$\frac{d}{d\varepsilon}S[\theta + \varepsilon\varphi]\bigg|_{\varepsilon=0} = ml\int_{t_1}^{t_2}\left[l\frac{d\theta}{dt}\frac{d\varphi}{dt} - \varphi g\sin\theta\right]dt = 0$$

We integrate the first term of the expression in parentheses in parts and taking into account that $\varphi(t) = 0$ at the moments t_1, t_2, we arrive at the following equation

$$ml \int_{t_1}^{t_2} \varphi \left[l \frac{d^2\theta}{dt^2} + d \sin\theta \right] dt = 0,$$

which, due to the arbitrariness of $\varphi(t)$, can be satisfied only if, for all $t_1 < t < t_2$,

$$\frac{d^2\theta}{dt^2} = -\frac{g}{l} \sin\theta$$

Thus, the law of motion of the mathematical pendulum is obtained using the Hamilton principle. Obviously, the same equation can be obtained using fundamental laws.

Indeed, the equation of oscillation of the pendulum is easy to obtain using the basic equation of the dynamics of rotational motion:

$$J \frac{d^2\theta}{dt^2} = -mgl \sin\theta$$

where $J = ml^2$ is the moment of inertia of the mathematical pendulum, and $M_1 = -mgl \sin\theta$ is the rotational moment created by gravity. Reducing the written equation by ml^2, we get the equation of oscillations, which does not depend on the mass of the load

$$\frac{d^2\theta}{dt^2} = -\frac{g}{l} \sin\theta$$

Note that the equation of oscillation of a mathematical pendulum, in contrast to the spring

$$m \frac{d^2r}{dt^2} = -kr, \quad t > 0$$

nonlinearly. This circumstance is associated with the more complex geometry of the rod-load system, namely the acceleration experienced by the load is disproportionate to the coordinate, as in the case of Hooke's law, but is a more complex function of the deviation from the equilibrium position (angle θ). If these deviations are small, then $\sin\theta \approx \theta$, and the model of small oscillations is linear

$$\frac{d^2\theta}{dt^2} = -\frac{g}{l}\theta$$

They are described by a formula similar to the formula for a spring pendulum,

$$r = A \sin(\omega t) + B \cos(\omega t)$$

where

$$\omega = \sqrt{\frac{g}{l}}$$

the natural frequency of small oscillations, and the quantities A, B are determined through $\theta(t = 0)$, $\frac{d\alpha}{dt}(t = 0)$.

Let us return to the question of constructing a phase portrait. Let us construct a phase portrait of a mathematical pendulum. For definiteness, we assume that the pendulum is a load of mass m, which can freely oscillate on a rigid suspension of length l. We write the total energy of the pendulum E when it is deflected by an arbitrary angle θ:

$$E = \frac{mL^2\omega^2}{2} + mgl(1 - \cos\theta)$$

We express ω from this equation:

$$\omega = \pm\sqrt{2E/(ml^2) - 2\omega_0^2(1 - \cos\theta)},$$

where $\omega_0 = \sqrt{g/l}$, then we can construct a phase portrait of the pendulum for arbitrary deflection angles.

We choose the initial values of the angular velocity of the pendulum from the range of $-8 < \omega_0 < 8$ rad/s in increments of 2 rad/s with zero initial deviation $\varphi_0 = 0$. It is clear that when the initial value of the angle is shifted by 2π, the type of phase portrait for oscillations does not change. Therefore, the graph can be immediately supplemented with phase trajectories for the cases $\varphi_0 = 2\pi$ and $\varphi_0 = -2\pi$.

It can be seen from the figure that when the initial value of the angular velocity modulo exceeds a certain number, the nature of the movement of the pendulum changes, namely the oscillations are replaced by rotation around the suspension point. Thus, there are two types of phase trajectories corresponding to two types of motion: closed trajectories (oscillations) and open paths (rotation around a suspension point).

We find a trajectory that will be a transition between the two types of motion. It corresponds to a certain initial value of the angular velocity, which can be found from the law of conservation of energy.

At the initial moment of time, the potential energy can be considered equal to zero (since the load is at the lowest point), and the total energy is equal to the kinetic:

$$E = E_k = \frac{mv_{x0}^2}{2} = \frac{m\omega_0^2 l^2}{2}$$

where $v_{x0} = \omega_0 l$—horizontal speed reported to the load at the initial moment of time.

The total energy at the highest point should be equal to the potential change (since we are looking for such an initial speed at which the pendulum stops at the highest point), i.e.,

$$E = \Delta E_{\mathrm{p}} = 2mg2l = 4mgl$$

Equating the obtained expressions, we find the initial value for the angular velocity corresponding to the transition trajectory:

$$\omega_0 = 2\sqrt{\frac{g}{l}} = 2\sqrt{\frac{9.81}{1}} \approx 6.264\,\text{rad/s}.$$

This speed corresponds to a phase curve called a separatrix. Shown in Fig. 3.7 is transition phase trajectory. The result is shown in Fig. 3.8.

Closed trajectories surround singular points of the "center" type with coordinates $\varphi = 2\pi n, \omega = 0$ (n is an integer). They correspond to the oscillations of the pendulum relative to a stable lower equilibrium position. Such oscillations occur if the energy of the system is $E < \frac{m\omega_0^2 l^2}{2} = 4mgl$. Moreover, if $E \ll 4mgl$, then the oscillations will be harmonic, and the phase trajectories will be ellipses (blue and yellow graphs of Fig. 3.8).

If $E \sim 2mgl$, then the oscillations will be inharmonic (green graph in Fig. 3.8). With increasing energy (amplitude of the pendulum), the period of oscillations will increase.

The upper equilibrium with coordinates $\varphi = (2n - 1)\pi$, $\omega = 0$ corresponds to singular points of the saddle type.

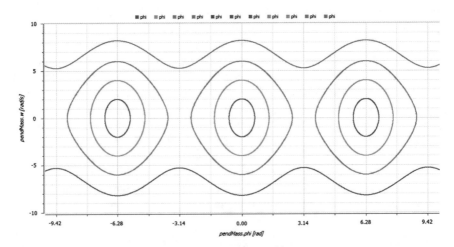

Fig. 3.7 Phase portrait of a mathematical pendulum at various initial values of the angular velocity ω_0 (in rad/s): blue—$\omega_0 = 2$ and $\omega_0 = -2$, yellow—$\omega_0 = 4$ and $\omega_0 = -4$, green—$\omega_0 = 6$ and $\omega_0 = -6$, red—$\omega_0 = 8$, purple—$\omega_0 = -8$

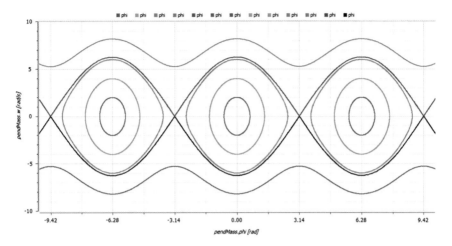

Fig. 3.8 Phase portrait of a mathematical pendulum at various initial values of the angular velocity ω_0 (in rad/s) with separatrices

Phase curves passing through unstable points—"saddles"—correspond to the energy $E = 4mgl$ and are called separatrices. They divide the phase space into regions with different behaviors. With an increase in the energy of the pendulum, its oscillations from quasiharmonic near points of the "center" type evolve to nonlinear periodic oscillations near the separatrices as indicated in Fig. 3.8. A further increase in energy leads to rotational motion.

Deviations of energy in one direction or another from the energy of motion along the separatrix lead to qualitatively different types of motion: oscillatory or rotational.

Thus, the separatrices divide the phase plane into two regions: the region of closed trajectories (oscillatory process) and the region of trajectories coming from infinity and going to infinity (rotation).

3.3 Hierarchical Principles of Model Building

Accounting for external disturbing forces

Let a known external force $F(r, t)$ acts on the load on the spring, depending on the time and position of the load. It can have a different nature, be generated by a gravitational field, have an electric or magnetic origin, etc. From Newton's second law, we get that, compared with the basic model of oscillations

$$m \frac{d^2 r}{dt^2} = -kr$$

an additional term appears on the right side of the equation:

$$m\frac{d^2r}{dt^2} = -kr + F(r,t)$$

In the simplest case, the applied force $F(r,t) = F_0$ is constant.

We carry out the change of variables: $\bar{r} = r - F_0/k$. We obtain for the new variable $\bar{r}$:

$$\frac{d^2\bar{r}}{dt^2} = -k\bar{r},$$

i.e., a constant force does not make changes in the oscillation process, with the exception that the coordinate of the neutral point at which the force acting on the load is zero is shifted by the value F_0/k. Thus, for example, a model of a vertical spring pendulum is constructed.

A much more complex picture of motion can arise when a time-dependent force $F(t)$ acts on the system. For definiteness, we consider the periodic external force $F(t) = F_0 \sin \omega_1 t$:

$$m\frac{d^2r}{dt^2} = -kr + F(t) = -kr + F_0 \sin \omega_1 t$$

The program code looks like in Fig. 3.9.

The coefficient f must be understood as the amplitude F_{max}.

Let us conduct an analytical and numerical analysis of the model. The solution of such a linear differential equation is found as the sum of the general solution of the corresponding homogeneous equation

$$r_{o.o.} = A \sin(\omega t) + B \cos(\omega t)$$

and particular solutions of the inhomogeneous equation

$$r_1(t) = C \sin \omega_1 t$$

```
forcedspringpendulum

model forcedspringpendulum
  parameter Real m(unit = "kg") = 1 "mass";
  parameter Real k(unit = "N/m") = 121 "coefficient of spring stiffness";
  parameter Real omega(unit = "Hz") = 11 "forcing frequency";
  parameter Real f(unit = "N") = 10 "amplitude of the force";
  Real x(unit = "m", start = 0.05) "displacement";
  Real v(unit = "m/s", start = 0.1) "velocity";
equation
  der(x) = v;
  m * der(v) = -k * x + f * cos(omega * time);
  ¤;
end forcedspringpendulum;
```

Fig. 3.9 WSM program code written in Modelica

where

$$C = \frac{F_0}{k - m\omega_1^2}$$

Knowing the spring oscillation frequency in the absence of external forces, or the natural frequency of the system $\omega = \sqrt{k/m}$, we obtain

$$C = \frac{F_0}{m\left(\omega^2 - \omega_1^2\right)}$$

As a result, for a general solution, we have

$$r(t) = A \sin \omega t + B \cos \omega t + \frac{F_0}{m\left(\omega^2 - \omega_1^2\right)} \sin \omega_1 t$$

So, the external force $F(t)$ leads not only to the appearance of additional oscillations in the system with frequency ω_1, but also to the possibility of resonance—to an unlimited increase in the amplitude of oscillations as $\omega_1 \to \omega$, as shown in Fig. 3.10.

Friction

Consider the result of the action of the frictional force on the system due to the resistance of the medium in which the load moves on the spring (air, water, etc.). In

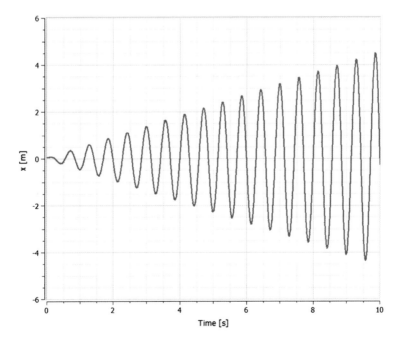

Fig. 3.10 Resonance phenomenon when a periodic pendulum acts on a spring pendulum

```
dampedspringpendulum
model dampedspringpendulum
  parameter Real m(unit = "kg") = 1 "mass";
  parameter Real k(unit = "N/m") = 100 "coefficient of spring stiffness";
  parameter Real gamma(unit = "N s/m") = 1 "viscosity";
  Real x(unit = "m", start = 0.05) "displacement";
  Real v(unit = "m/s", start = 0.1) "velocity";
equation
  der(x) = v;
  m * der(v) = -k * x - gamma * v;
  □;
end dampedspringpendulum;
```

Fig. 3.11 WSM program code written in Modelica

this case, the friction force is not constant, but substantially depends on the speed of movement. This dependence is described by the well-known Stokes' formula

$$F = -\mu v = -\mu \frac{dr}{dt},$$

where the coefficient $\mu > 0$ is determined by the dimensions of the load, the density of the medium, its viscosity, etc. The equation of motion in a viscous medium has the form

$$m \frac{d^2 r}{dt^2} = -kr + F(v) = -kr - \mu \frac{dr}{dt}$$

The program code looks like in Fig. 3.11.

We will carry out analytical and numerical analysis of the model.

We find the general solution of this linear equation, having previously got rid of the term with the first derivative. Substitution in the replacement equation $r(t) = \bar{r}(t)e^{\alpha t}$ gives the equation $\bar{r}(t)$ for the new function

$$m \left(e^{\alpha t} \frac{d^2 \bar{r}}{dt^2} + \alpha e^{\alpha t} \frac{d\bar{r}}{dt} + \alpha e^{\alpha t} \frac{d\bar{r}}{dt} + \alpha^2 e^{\alpha t} \bar{r} \right) = -k\bar{r}e^{\alpha t} - \mu e^{\alpha t} \frac{d\bar{r}}{dt} - \mu \alpha e^{\alpha t} \bar{r}$$

Reducing the factor $e^{\alpha t}$ in it and setting $\alpha = -\mu/(2m)$, we arrive at the equation

$$m \frac{d^2 \bar{r}}{dt^2} = -\left(k - \frac{\mu^2}{4m} \right) \bar{r} = -k_1 \bar{r}$$

In contrast to the equation for forced oscillations, the first factor on the right-hand side of the obtained expression can change sign depending on the values of the parameters k, μ, and m of the system, which, taking into account the relation $r(t) = \bar{r}(t)e^{\alpha t}$, leads to its other behavior with respect to the standard case of an ideal harmonic oscillator.

At low viscosity, i.e., for $k - \mu^2/(4m) = k_1 > 0$, the solution $\bar{r}$ is given by the formula $r = A\sin(\omega t) + B\cos(\omega t)$, and for $r(t)$, we have

$$r = \bar{r}e^{\alpha t} = e^{-t\mu/(2m)}(A\sin(\omega t) + B\cos(\omega t))$$

where the constants are from the initial conditions. Oscillations damping with time with a frequency ω occur in the system, as shown in Fig. 3.12.

If $k_1 = 0$, then the quantity $(d\bar{r}/dt$ is constant, which means that $\bar{r}(t) = ct + c_1$. For $r(t)$, taking into account the initial conditions, we obtain

$$r = e^{-t\mu/(2m)}\left(\left(v_0 + \frac{\mu r_0}{2m}\right)t + r_0\right)$$

In this case, there are no vibrations due to the overwhelming effect of viscous friction forces, as shown in Fig. 3.13. The system can only pass the point $r = 0$ once, for which it is necessary and sufficient that the conditions $v_0 < -\mu r_0/(2m)$, $r_0 > 0$

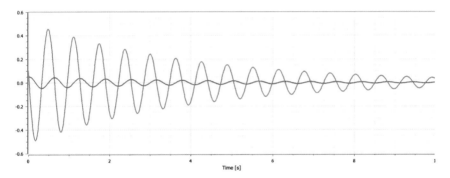

Fig. 3.12 Graphs of the dependence of displacement and velocity on the time of a spring pendulum in a viscous medium ($m = 1$ kg, $k = 100$ N/m, $\gamma = 0.5$ N s/m)

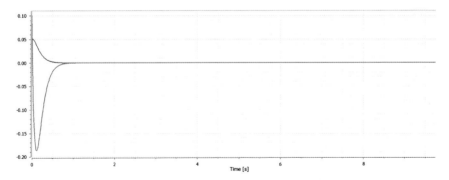

Fig. 3.13 Graphs of the displacement and velocity versus time of a spring pendulum in a viscous medium ($m = 1$ kg, $k = 100$ N/m, $\gamma = 20$ N s/m)

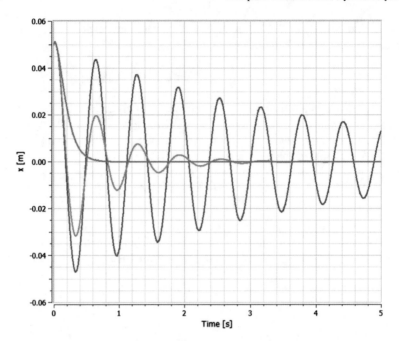

Fig. 3.14 Comparison of load displacements in media with different viscosities (blue line—low viscosity, orange line—high viscosity, green line—overwhelming viscosity)

or $v_0 > -\mu r_0/(2m)$, $r_0 < 0$, i.e., the initial velocity of the ball should be sufficiently large and directed to the point $r = 0$. In this case, obviously, the velocity of the ball $v(t) = dr/dt$ can change sign only once.

Finally, at high viscosity, the action of the friction force is so significant, as shown in Fig. 3.14, that for any r_0, v_0 the ball "gets stuck" in the medium, never passing the point $r = 0$, but only approaching it unilaterally as $t \to \infty$. Indeed, for $k_1 < 0$, the solution of the equation is constant (the assumption of otherwise immediately leads to a contradiction with the equation); therefore, the quantity $r(t)$ also does not change sign. The behavior of the function $\bar{r}(t)$ as $t \to \infty$ can be understood from the properties of the first integral of the equation

$$m\left(\frac{d\bar{r}}{dt}\right)^2 = -k_1\bar{r}^2 + \text{const},$$

which is easy to obtain by multiplying both sides of the equation by $d\bar{r}/dt$ and integrating once over t. Assumptions that $\bar{r}(t) \to \infty$ or $\bar{r}(t) \to C_1 \neq 0$ as $t \to \infty$, and thus $r(t) \to 0$, $t \to \infty$.

So, the movement of the system in a viscous medium is characterized by a large variety with respect to the ideal situation, and in all cases, it occurs with damping.

Let us again turn to the question of constructing a phase portrait.

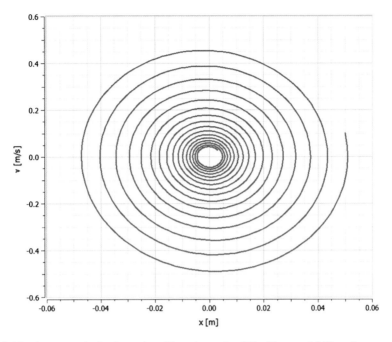

Fig. 3.15 Phase portrait of a damped oscillator ($\omega_0 = 2$ rad/s) with a special "focus"-type point

In Fig. 3.15, a singular point of the center type was shown. It is characteristic of undamped oscillations near the equilibrium position. If there is damping, each ellipse becomes a spiral (Fig. 3.15), and a singular point at the origin becomes the focus. For a family of phase trajectories, the focus is an attractor: All phase trajectories, regardless of where they start, asymptotically approach the focus, making around it an infinite number of revolutions along increasingly contracting turns. If the damping is weak, then the spiral consists of a large number of closely spaced turns. The stronger the damping, the farther the turns are from each other.

With very strong damping, the phase portrait also changes qualitatively, taking the form shown in Fig. 3.16. Here, the origin is also a singular point, but of a different type. The attractor of phase trajectories from the focus turns into a special point such as a knot: All phase trajectories of non-oscillatory movements approach this node directly without winding, without completing one revolution. At the node, all phase trajectories are tangent to the inclined line passing through it, and along this line are contracted to a singular point.

Recall that when considering a mathematical pendulum, we came across another type of singular points that are possible in nonlinear conservative systems, namely a singular point of the saddle type through which two phase trajectories pass.

So, by the type of phase trajectories surrounding the singular points, the following types of these points are distinguished: center, focus, knot, and saddle. These concepts, borrowed from the theory of differential equations, are very useful for

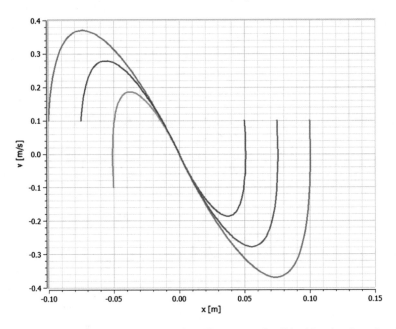

Fig. 3.16 Phase portrait of a strongly damped oscillator ($\omega_0 = 2$ rad/s) with a singular point of the "node" type at the origin

describing the behavior of a dynamical system. A special point is reached only after an infinitely long time.

Comment

The same complication, taking into account the force of friction, can be considered for the model of a mathematical pendulum. We complicate the task and take into account that when moving in a viscous medium, the pendulum experiences the action of a resistance force. We write down the rotational moments created by gravity ($M_1 = -mgl \sin\theta$) and resistance force ($M_1 = -bl^2 \frac{d\theta}{dt}$), under the assumption that the resistance force is directly proportional to the speed:

$$J\frac{d^2\theta}{dt^2} = -mgl \sin\theta - bl^2 \frac{d\theta}{dt}$$

As a result, the equation of free damped oscillations of a mathematical pendulum is obtained:

$$\frac{d^2\theta}{dt^2} = -\frac{g}{l} \sin\theta - \frac{b}{m}\frac{d\theta}{dt}$$

The friction in the system causes damping of the oscillations and the value of b/m characterizes the speed of this damping. If friction is negligible, then in the case of small deviations $\sin\theta \approx \theta$, this equation goes over into the equation of free undamped oscillations, the solution of which is a periodic function.

Dry friction

Dry or Coulomb friction is observed when solids are in contact and move relative to one another at the point of contact. Friction forces in the absence of lubrication are almost independent of the magnitude of the speed of movement; their direction is opposite to the speed of relative displacement.

In this case, the friction force is $F = k_1 P$, where k_1 is the coefficient of friction, $P = mg$ is the weight of the ball. It is always directed against the movement of the ball, its sign is opposite to the sign of the speed of the ball $v = dr/dt$, i.e.,

$$F = -k_1 mg\,\text{sign}(dr/dt)$$

The motion of the ball obeys the equation

$$m\frac{d^2r}{dt^2} = -kr - k_1 mg\,\text{sign}\left(\frac{dr}{dt}\right)$$

which looks like the equation of forced oscillations with constant force $F(r, t) = F_0$. However, due to the alternating force, it does not reduce to the standard equation of oscillations. This circumstance serves as an expression of the fact that the equations of forced oscillations and the equation of motion under the action of dry friction describe essentially different processes. In particular, the amplitude of the load oscillations in the latter case decreases with time. This can be easily verified by rewriting the last equation in the form

$$m\frac{dv}{dt} + kr = -k_1 mg\,\text{sign}\,v,$$

multiplying both sides of this expression by $v/2$ and taking into account the fact that $v = dr/dt$, we obtain

$$\frac{m}{2}\frac{dv^2}{dt} + \frac{k}{2}\frac{dr^2}{dt} = -\frac{1}{2}k_1 mg\,\text{sign}\,v * v$$

Taking into account that the sum of the kinetic and potential energy of the system $E(t) = E_k(t) + E_p(t)$, is on the left side of the last equality under the sign of the derivative, and the right side of the expression is negative for $v \neq 0$, we have

$$\frac{dE(t)}{dt} < 0, \quad v \neq 0$$

$$\frac{dE(t)}{dt} = 0, \quad v = 0$$

i.e., the total energy $E(t)$ decreases with time. Since at the moments when the load reaches its maximum amplitude $|r_m(t)|$, its speed and kinetic energy E_k are equal to zero, then at these moments

$$E_p = -kr_m^2(t)/2 = E(t)$$

and since $E(t)$ decreases, the amplitude $|r_m(t)|$ is also a decreasing function of time.

$$E_k + E_p = E_0 - rx = \overline{E_0}$$

Points of change of direction are characterized by the values $v = 0$, or $E_k = 0$. The corresponding values of the amplitude x are obtained as the points of intersection of the E_p curve with the "direct energy loss" $E_0 - rx$. If, for example, the movement begins at $x = x_1 < 0$ and $v = 0$, then the first point of change in the direction of motion is obtained at the amplitude $x = x_2 > 0$ and we leave the region $v > 0$. To move in the opposite direction, you need to substitute another value $E_0 = E_{02}$ and the opposite sign in front of r in the equation. Thus, a direct loss of energy with a positive slope is obtained, which, of course, must dock with a straight line corresponding to the first oscillation, i.e., go through the intersection point of the first line with the E_p curve for $x = x_2$. Another point of intersection of the second straight line with the E_p curve gives the next point of change in the direction of motion and, accordingly, the amplitude $x = x_3$.

You can continue this construction and find a sequence of points of change of direction. The sequence x_n ends when the slope of the E_p curve becomes less than the slope of the direct energy loss.

This can be easily explained physically: As the deviation x decreases, the restoring force decreases, while the friction force remains constant. Starting from a certain value of the deviation x, the friction force becomes larger than the restoring force, and the restoring force cannot cause the oscillator to shift from the corresponding point of change of direction. The vibrations end in the dead zone, determined by the value of the friction force r.

From the point of view of the theory of dynamical systems, the system "spring pendulum with dry friction" should be attributed to hybrid systems.

In practice, one often has to deal with discrete–continuous models of dynamic systems, which are called hybrid systems. Other names for such systems are "variable structure systems" and "event-driven systems." The variable structure of the models is due to the presence of slow (continuous) and fast (discrete) processes in the system. Thus, a distinctive feature of the complex behavior of a hybrid system is the multitude of qualitatively different and successively changing modes of functioning. Continuous behavior is called the state of the hybrid system, while switching of the states—discrete events. The execution time of discrete events is not taken into account, from the point of view of the functioning of the system they are performed instantly.

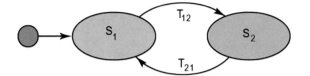

Fig. 3.17 Dry friction spring pendulum state diagram

Hybrid models allow systems to be described in a variety of engineering applications. So, for example, in mechanical objects, continuous movement can be interrupted or corrected by some physical effect. So, in the system "spring pendulum with dry friction," a continuous state is the movement of the spring pendulum in one direction. The stopping moment is a discrete event, after which the system begins to move in a new continuous mode of the oscillatory process (in the opposite direction).

Hybrid mathematical models at the development stage are conveniently described graphically using state diagrams (UML standard). The state diagram is a directed graph whose vertices correspond to the states of the system under study, and the edges describe discrete events, that is, transitions from one state to another. There are also two service states: initial and final. The initial state is present in any state diagram; the final state may or may not depend on the given system.

Let us describe the mathematical model of this system in the form of a state diagram (Fig. 3.17). The black circle in the figure indicates the initial state, S_1 and S_2 are the states (continuous) of the model, T_{12} is the transition from S_1 to S_2.

We plot the displacement and speed of the load. This graph is an alternation of pieces of sinusoids. For each cycle of oscillations, the maximum deviation decreases by the same amount—twice the width of the stagnation zone—as shown in Fig. 3.18.

This means that with dry friction, successive maximum deviations decrease in arithmetic progression, linearly, as opposed to an infinitely long decrease in geometric progression with viscous friction.

From the energy relations, one can also derive the equations of phase trajectories:

$$v = +\sqrt{\frac{2}{m}(E_0 - rx - E_p)}, \quad v > 0$$

$$v = -\sqrt{\frac{2}{m}(E_0 + rx - E_p)}, \quad v < 0$$

Accordingly, expressions are obtained for the oscillation period. At the same time, the time intervals for which the vibrations are made are calculated separately. In this way,

$$T = T_1 + T_2$$

When the restoring force is linear (Hooke's law), the energy ratio takes the form

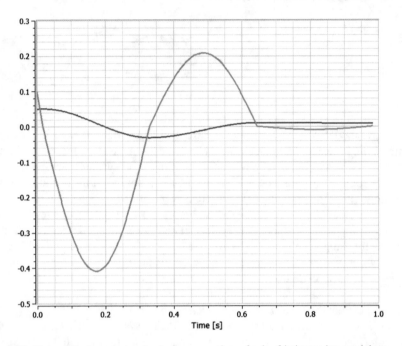

Fig. 3.18 Plots of displacement and velocity versus time of a dry friction spring pendulum

$$\frac{1}{2}mv^2 + \frac{1}{2}cx^2 + rx = E_0$$

$$\frac{1}{2}mv^2 + \frac{1}{2}c\left(x + \frac{r}{c}\right)^2 = E_0 + \frac{r^2}{2c} = E_{\dot{0}}$$

or at $\omega_0^2 = c/m$

$$\left(\frac{v}{\omega_0}\right)^2 + \left(x + \frac{r}{c}\right)^2 = \frac{2E_{\dot{0}}}{c}$$

If the phase trajectories described by this relation are constructed in the phase plane, then we obtain circles whose center is shifted from the origin along the abscissa to the left by r/c (Fig. 3.19).

This center on the left is the center of all the semicircles in the upper half-plane. Correspondingly, the center of all semicircles in the lower half-plane is located on the right. Phase trajectories are composed of a sequence of semicircles of this kind, which, when crossing the abscissa axis, always go one into another. From the phase portrait, it is also easy to see that the motion should come to a state of rest through a finite number of vibrations. With each half-oscillation, a decrease in amplitude occurs, equal to

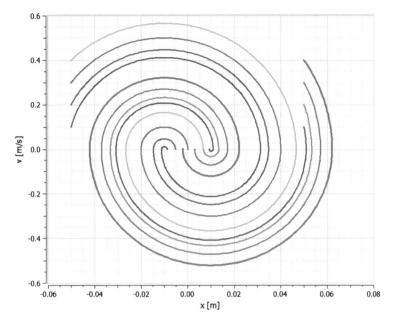

Fig. 3.19 Phase portrait of an oscillator with dry friction and linear regenerative force

$$\Delta x = 2r/c$$

Analysis of nonlinear systems by the example of the occurrence of nonlinear deformation in a spring pendulum

A spring pendulum with nonlinear deformation is a mechanical system consisting of a spring in which, during deformation, a force arises nonlinearly associated with the extension of the spring, one end of which is rigidly fixed, and the second is attached a load in the form of a material point with a given mass m.

This means that Hooke's law is not satisfied for the deformation of the spring. Oscillation equation

$$m\frac{d^2r}{dt^2} = -k(r)r$$

where the function $k(r) > 0$ describes the spring stiffness—one of the relatively few nonlinear equations for which a general solution can be written. Introducing the velocity $\upsilon = \frac{dr}{dt}$, we rewrite the last equation in the form

$$m\frac{d\upsilon}{dt} = -k(r)r$$

$$\upsilon = \frac{dr}{dt}$$

Dividing the first of these equations by the second, we obtain a first-order nonlinear equation

$$m\frac{d\upsilon}{dr} = \frac{-k(r)r}{\upsilon}$$

Separating the variables in the resulting equation

$$m\upsilon d\upsilon = -k(r)r dr$$

And integrating the last equation twice, we find

$$\upsilon^2 = \left(\frac{dr}{dt}\right)^2 = -2\int_0^r \frac{k(r')}{m}r'dr' + C$$

$$\frac{dr}{dt} = \pm\sqrt{C - 2\int_0^r \frac{k(r')}{m}r'dr'}$$

$$t = \pm\int_0^r d\bar{r}\left(\sqrt{C - 2\int_0^r \frac{k(r')}{m}r'dr'}\right)^{-1} + C_1$$

where in the implicitly written general solution, the constants C and C_1 can be determined, knowing the initial data.

As we already know, in the linear case $k(r) = $ const, the phase trajectories of the system are concentric ellipses with a center at the origin, the main semi-axes of which are determined by the initial energy of the system and the "motion" along which describes the oscillation process that is periodic in time.

We now consider a strongly nonlinear system in which the spring behaves as "super soft," for example, $k(r) = 1/(\alpha + r^2)$, $\alpha > 0$. In the limiting case $\alpha = 0$, the nonlinear equation takes the form

$$m\frac{d\upsilon}{dt} = -\frac{1}{r}$$

$$\upsilon = \frac{dr}{dt}$$

And its solution is fundamentally different from the solution for a harmonic oscillator (Fig. 3.20a), in that the energy is not conserved; moreover, it grows unlimitedly as $r \to \pm 0$ (Fig. 3.20b). With the weakening of nonlinearity, the process becomes normal.

Consider another type of nonlinearity. We assume that the elastic force is related to the elongation r by the ratio:

(a) **(b)**

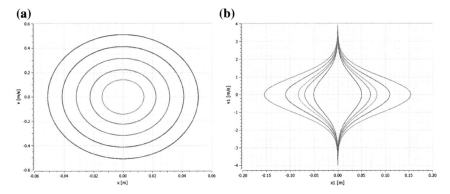

Fig. 3.20 **a** Phase portrait of a harmonic oscillator. **b** Phase portrait of a nonlinear oscillator with a "super soft" spring

$$F_{ela} = k(1 + \alpha r + \cdots)r$$

Those, for spring stiffness, we will take into account only the first two terms in the expansion in a Taylor series in the vicinity of the equilibrium position. When removing this system from the equilibrium position, it will begin to perform non-harmonic oscillations, slightly differing from harmonic.

Based on Newton's second law and a given type of elastic force, the behavior of a spring pendulum with nonlinear deformation is described by the equation:

$$m\frac{d^2x}{dt^2} = -k\left(x + \alpha x^2\right)$$

If the vibrations occur in a viscous medium, then the equation of motion will have the form:

$$m\frac{d^2x}{dt^2} = -k\left(x + \alpha x^2\right) - \gamma\frac{dx}{dt}$$

A general Text View recorded in Modelica is shown in Fig. 3.21. Please note that the viscosity of the medium in the first experiment is zero; however, by performing a series of experiments in the Simulation Center, we will be able to change this value in the experiment settings in the future.

As a result of the experiment, we can observe the slightly inharmonic nature of the displacement and speed of the load, as shown in Fig. 3.22.

When constructing the phase portrait, a series of experiments was carried out where the internal graph, which is an ellipse, corresponds to the absence of a non-linear component. When the coefficient is "turned on" and further increases with the nonlinear term, we observe a gradual deformation of the ellipse, as shown in Fig. 3.23.

nonleneardampedspringpendulum

```
model nonleneardampedspringpendulum
   parameter Real m(unit = "kg") = 1 "mass";
   parameter Real k(unit = "N/m") = 100 "coefficient of spring stiffness";
   parameter Real gamma(unit = "N s/m") = 0 "viscosity";
   parameter Real alpha = 1 "nonlinear cofficient";
   Real x(unit = "m", start = 0.05) "displacement";
   Real v(unit = "m/s", start = 0.1) "velocity";
equation
   der(x) = v;
   m * der(v) = (-k * (x + alpha * x * x)) - gamma * v;
   ▫;
end nonleneardampedspringpendulum;
```

Fig. 3.21 Nonlinear spring force model

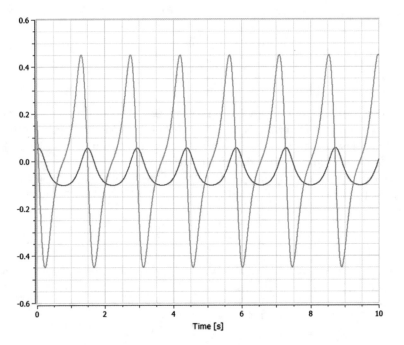

Fig. 3.22 Characteristics of a spring pendulum with nonlinear elastic force

Motion in non-inertial reference systems

Resonance in the system can be caused due to the action of inertial forces. Consider the spring pendulum again. Let the spring attachment point move according to a given law $r_0(t) = f(t)$. Then, in the coordinate system associated with this point, in addition to the spring tension, the inertia force is equal to $ma(t)$, where $a(t)$ is the acceleration due to the movement of the coordinate system, $a(t) = \mathrm{d}^2 f/\mathrm{d}t^2$. In this coordinate system, the movement of the load is described by the equation

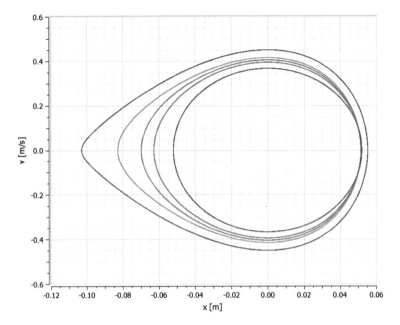

Fig. 3.23 Phase portrait with nonlinear elastic force

$$m\frac{d^2 r}{dt^2} = -kr + F(t)$$

where

$$F(t) = -ma(t) = -\frac{md^2 f}{dt^2}$$

where $F(t)$ is a given function of time. As in the case of a driving force, with a corresponding periodic movement of the attachment point in the system, resonance occurs.

With a more complex geometry, the inertia forces of the system can depend not only on time t, but also on the coordinate r. If the spring is worn on a rod moving with an angular velocity $\omega(t)$, then the centrifugal inertia is

$$F = m\upsilon^2(t)/R$$

where

$$\upsilon(t) = \omega(t)R$$

$R = R_0 + r$, R_0—spring length in unloaded condition, r—deviation of the load from the neutral position, $r > -R_0$.

The equation of motion of the load takes the form

$$m\frac{d^2r}{dt^2} = -kr + F(r, t)$$

where

$$F(r, t) = m\omega^2(t)(R_0 + r),$$

or

$$m\frac{d^2r}{dt^2} = -\big(k - m\omega^2(t)\big)r + m\omega^2(t)R_0,$$

and, obviously, when $r \ll R_0$, the written linear equation goes over into the equation with an external exciting force depending on the time $F(t) = -m\omega^2(t)R_0$. However, in this case, resonance is impossible, since the external force is always directed in one direction and is not able to swing the system.

As an example, illustrating motion in non-inertial reference frames, we consider the so-called elliptical pendulum.

A simple mathematical pendulum is mounted on a trolley of mass m_1 moving without friction along a smooth rail. The trolley moves with constant transport acceleration a_0, and a pendulum with a load of mass m_2 at the initial moment of time hangs vertically down on an inextensible string of length l. After the start of the movement of the trolley, the pendulum enters into an oscillatory process. It is necessary to build a mathematical model of the "trolley–pendulum" system and experimentally investigate the dependence of the kinematic characteristics of the pendulum on the acceleration of the trolley.

When the mechanical system moves, the tension of the thread changes, which leads to a change in the restoring force, and, consequently, in the frequency and period of oscillations of the pendulum.

The equation of motion of a material point in a non-inertial reference frame without rotation can be represented as:

$$\sum F = m(a_0 + a_{\text{rel}}),$$

where a_{rel} is the acceleration of the body relative to the non-inertial reference frame and a_0 is the portable acceleration of the body. Thus, the acceleration of the load a consists of a_{rel}—acceleration of the load relative to the trolley and a_0—acceleration of the trolley itself:

$$a = a_0 + a_{\text{rel}}$$

Load speed can also be represented as a decomposition into relative and portable speeds.

We construct a model of this system in generalized coordinates using the Lagrange equations of the second kind. The second-order Lagrange equations are a system of ordinary second-order differential equations. They describe the movement of a mechanical system subordinate to ideal bonds. Lagrange equations of the second kind can be used to study the motion of any mechanical system with geometric connections, regardless of how many points or bodies enter the system, how the bodies move, and what kind of movement is considered.

If the motion of the holonomic system is described by the generalized coordinates q and the generalized velocities $\dot{q}$, then the equations of motion have the form

$$\frac{d}{dt}\left(\frac{\partial T(q,\dot{q})}{\partial \dot{q}}\right) - \frac{\partial T(q,\dot{q})}{\partial q} = Q_{\text{ext}}$$

where $T(q,\dot{q})$ is the kinetic energy of the system and Q_{ext} is the generalized force. The difference in the total time derivative of the partial derivative of kinetic energy with respect to the generalized velocity and the partial derivative of kinetic energy with respect to the generalized coordinate is equal to the generalized force. We also note that if the dimension q is length, then Q_{ext} has the dimension of ordinary force; if the generalized coordinate q is an angle (a dimensionless dimension), then Q_{ext} has the dimension of the moment of force.

So, in order to write the Lagrange equation, you must sequentially perform the following steps:

- set generalized coordinates;
- record kinetic and potential energy: $T(q,\dot{q})$ and $W(q)$;
- calculate Lagrangian $L(q,\dot{q})$;
- identify external applied forces Q_{ext};
- use $L(q,\dot{q})$ and Q_{ext} to get the expression for the Lagrange equation.

The trolley–pendulum system has two degrees of freedom and is located in the field of gravity. Let $q_1 = y$ as the generalized coordinates be the offset of the trolley from the origin, and $q_2 = \theta$ is the angle of deviation of the rod from the vertical, as shown in Fig. 3.24.

We represent the kinetic energy of the system as the sum of the kinetic energy of the trolley and pendulum:

$$T = T_1 + T_2 = \frac{1}{2}m_1\dot{y}^2 + \frac{1}{2}m_2 v_2^2,$$

here v_2 is the absolute speed of the load

$$\overrightarrow{v_2} = \vec{v}^{\text{rel}} + \vec{v}^{\text{abs}}$$

Given that

$$v^{\text{rel}} = l\dot{\theta}, \quad v^{\text{abs}} = \dot{y}$$

Fig. 3.24 Dynamic system: elliptical pendulum

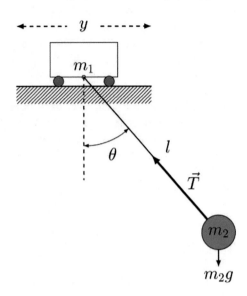

we find

$$v^2 = l^2\dot{\theta}^2 + \dot{y}^2 + 2l\dot{\theta}\dot{y}\cos\theta$$

Then for kinetic energy, we get:

$$T = T_1 + T_2 = \frac{1}{2}(m_1 + m_2)\dot{y}^2 + \frac{1}{2}m_2 l^2\dot{\theta}^2 + m_2 l\dot{\theta}\dot{y}\cos\theta$$

Potential energy can be calculated as

$$W(q) = -m_2 gl\cos\theta$$

Thus, the Lagrangian of the system will have the form

$$L = \frac{1}{2}(m_1 + m_2)\dot{y}^2 + \frac{1}{2}m_2 l^2\dot{\theta}^2 + m_2 l(\dot{\theta}\dot{y} + g)\cos\theta$$

We compose the first Lagrange equation:

$$\frac{\partial L}{\partial \dot{y}} = (m_1 + m_2)\dot{y} + m_2 l\dot{\theta}\cos\theta$$

$$\frac{\partial L}{\partial y} = 0$$

So,

$$\frac{d}{dt}\frac{\partial L}{\partial \dot{y}} = \frac{d}{dt}\left[(m_1 + m_2)\dot{y} + m_2 l\dot{\theta}\cos\theta\right] = 0$$

We compose the second Lagrange equation:

$$\frac{\partial L}{\partial \dot{\theta}} = m_2 l^2\dot{\theta} + m_2 l\dot{y}\cos\theta$$

$$\frac{\partial L}{\partial \theta} = -m_2 l(\dot{\theta}\dot{y} + g)\sin\theta$$

$$\frac{d}{dt}\frac{\partial L}{\partial \dot{\theta}} = m_2 l^2\ddot{\theta} + m_2 l\ddot{y}\cos\theta - m_2 l\dot{\theta}\dot{y}\sin\theta$$

So,

$$m_2 l^2\ddot{\theta} + m_2 l\ddot{y}\cos\theta + m_2 lg\sin\theta = 0$$

As a result, we obtain a system of equations describing the motion of the system under consideration.

$$\frac{d}{dt}\left[(m_1 + m_2)\dot{y} + m_2 l\dot{\theta}\cos\theta\right] = 0$$

$$l\ddot{\theta} + \ddot{y}\cos\theta + g\sin\theta = 0$$

Simulation of the constructed model at various accelerations of the cart gives us the following series of graphs, as shown in Fig. 3.25. The displacement graph with the smallest amplitude (yellow graph) corresponds to a resting cart $a_0 = 0$. With increasing acceleration of the cart, the amplitude of the oscillations increases; however, the oscillations themselves remain close to harmonic.

Physical complications that do not lead to more complex system behavior

Note that geometry, which is more complicated than the initial case, does not always mean more complex object behavior.

Consider, for example, a load attached to two springs with stiffness k_1 and k_2. The connection shown in Fig. 3.26 is called parallel.

We place the origin at the point where the forces acting on the load from the side of both springs balance each other (in this case, some condition on the system parameters must be observed so that the load cannot touch one of the attachment points). According to Hooke's law, when r is deflected, the force acts on the load from the left spring side—$k_1 r$, and from the right side—$k_2 r$ (both forces are directed in one direction, since when the first spring is stretched, the second spring, on the contrary, is compressed). As a result, we arrive at the same equation as in the case of a single spring

$$m\frac{d^2 r}{dt^2} = -k_1 r - k_2 r = -kr$$

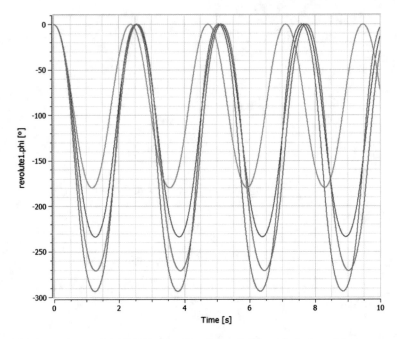

Fig. 3.25 Dependence of the amplitude of oscillations of an elliptical pendulum on portable acceleration (a graph of the angular displacement from the horizontal is shown $\varphi = \theta - \frac{\pi}{2}$)

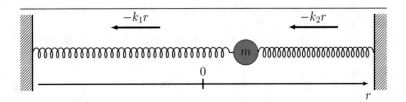

Fig. 3.26 Parallel spring connection

but with increased stiffness $k = k_1 + k_2$, consisting of the stiffnesses of both springs. So, Fig. 3.22 shows the oscillation graphs of two spring pendulums with individual stiffness $k_1 = 0.4$ kg/s^2 (yellow line) and $k_2 = 0.6$ kg/s^2 (green line) and the oscillation graph when they are connected in parallel. It is easy to see that the resulting oscillation is equivalent to the movement of a single spring with rigidity $k = 1$ kg/s^2 (Fig. 3.27).

If you assemble the system shown in Fig. 3.28, then we get a series connection of springs.

When removing the body from the equilibrium position, the elongations of the springs x_1 and x_2 are different. However, the deformations of the springs are not arbitrary: Their sum is always equal to the displacement of the body

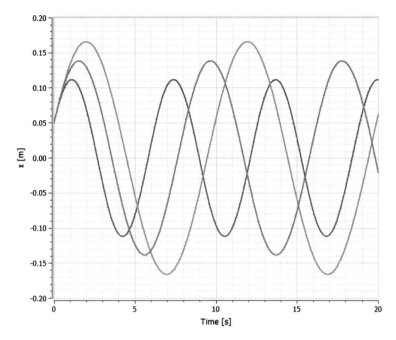

Fig. 3.27 Graphs of the displacements of individual springs and a graph of their resulting displacement with parallel connection

Fig. 3.28 Series connection of springs

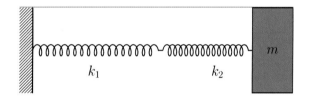

$$x_1 + x_2 = x$$

The elastic forces based on the assumption of inertia of the springs should be equal. Therefore, we obtain:

$$x = -\frac{F_{\text{ela}}}{k_1} - \frac{F_{\text{ela}}}{k_2} = -F_{\text{ela}}\frac{k_1 + k_2}{k_1 k_2}$$

where

$$k = \frac{k_1 + k_2}{k_1 k_2}$$

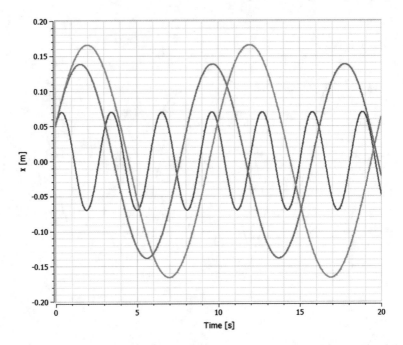

Fig. 3.29 Graphs of the displacements of individual springs and a graph of their resulting displacement in series connection

is rigidity of an equivalent spring capable of replacing two real springs connected "in series." The differential equation of motion of the body also does not contain any complicating elements:

$$m\frac{\mathrm{d}^2 x}{\mathrm{d}t^2} = -\frac{k_1 + k_2}{k_1 k_2} x$$

Similarly, to the previous experiment, we take two springs with stiffness $k_1 = 0.4$ kg/s^2 (yellow line) and $k_2 = 0.6$ kg/s^2 (green line) and connect them in series. Get a blue graph. It coincides with the spring displacement schedule with stiffness $k = 4.17$ kg/s^2, which is consistent with the theoretical formula, as shown in Fig. 3.29.

3.4   Universality of a Computer Model and Equivalent Physical Systems

Any dynamic system corresponds to some physical model. It may turn out that the same physical (or mathematical) model is used to describe systems that are completely different in nature. Let us consider this case as an example of oscillatory systems [8].

Fig. 3.30 Torsion spring
oscillator

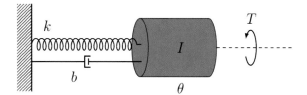

To describe real vibrational dynamical systems, some physical model is used, which is called a linear or harmonic oscillator. This model introduces some basic assumptions.

If the oscillator is once removed from the state of equilibrium, then a restoring force will arise in such a system, which will tend to return the oscillator back to the equilibrium position. The restoring force in magnitude is directly proportional to the displacement and directed against the displacement. The physical quantities that describe the oscillations of the oscillator change with time according to the harmonic law of cosine or sine.

However, there are quite a lot of oscillatory systems to which the linear oscillator model is applicable:

(a) spring pendulum—a load suspended on an elastic spring, which obeys Hooke's law at small displacements;
(b) the mathematical and physical pendulums that make small deviations from the vertical in the gravitational field;
(c) an electric oscillatory circuit, consisting of a precharged capacitor, coil, and resistor, which are connected in series in a circuit;
(d) torsion spring oscillator.

Let us consider it in more detail; see Fig. 3.30.

The oscillator consists of a disk (or rotor), which can rotate relative to a fixed axis perpendicular to the plane in which the rotor is located. The rotor is engaged with a weightless coil spring at one of its ends. The other end of the spring is rigidly fixed. When the disk is rotated by a certain angle relative to point 0 (equilibrium position), the spring, elastically deformed, twists, and the rotor stops in the extreme position. Then the spring unwinds, returning the disk to the equilibrium position. The rotor passes this position and deviates in the other direction to a stop. The spring makes it again return to the equilibrium position, etc. Thus, the torsion oscillator performs torsional vibrations from left to right relative to point 0.

An external torque T may be applied to the disk.

The physical parameters of the system characterizing the torsion oscillator: the moment of inertia of the disk I, the spring stiffness k (torsion modulus), the damping constant b (taking into account the viscous friction of the medium).

When the disk is rotated from the equilibrium position by an angle θ, a spiral spring attached to it (the other end of which is fixed motionless) creates a returning moment N proportional to the deflection angle:

$$N = -k\theta,$$

The proportionality coefficient k is called the spring stiffness. Applying the basic equation of the dynamics of rotation of a rigid body around a fixed axis to the motion of a disk with an inertia moment I,

$$N = I\frac{d\omega}{dt} = I\frac{d^2\theta}{dt^2}$$

we obtain the following differential equation of the natural oscillations of the torsion spring oscillator:

$$I\frac{d^2\theta}{dt^2} = -k\theta$$

or

$$\frac{d^2\theta}{dt^2} + \frac{k}{I}\theta = 0$$

The general solution of this equation is a simple harmonic oscillation. The oscillations occur with an angular frequency ω_0, the square of which is proportional to the spring stiffness k and inversely proportional to the moment of inertia I of the disk, i.e. this mathematical model is equivalent to the spring pendulum model, but with other physical parameters.

In the presence of a viscous friction force, the braking moment of this force proportional to the angular velocity of the disk will be present in the differential equation $\dot{\theta}$:

$$I\ddot{\theta} + b\dot{\theta} + k\theta = 0$$

And again, they got a model similar to the spring pendulum model with damping. The graphs characterizing the kinematic and dynamic behavior of such a model are already known to us—they will coincide with the graphs for a spring pendulum with the difference that instead of linear displacement and velocity, angular displacement and angular velocity will act as kinematic characteristics.

Apply external torque to the disk:

$$I\ddot{\theta} + b\dot{\theta} + k\theta = T$$

Thus, we will be able to additionally investigate the constant effect on the system, which we have not studied above.

Let the system consists of a disk with an inertia moment of $I = 0.5$ kg m^2 and a spring of rigidity $k = 0.1$ kg/s^2 with a fixed end. Let the system be at rest at the initial

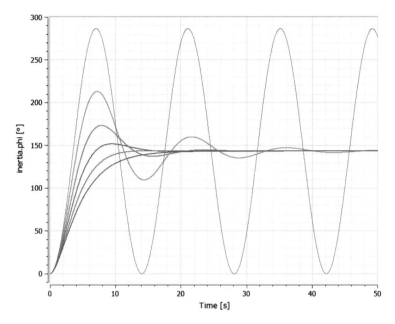

Fig. 3.31 Graph of the angular displacement of the disk at different viscosity of the medium

instant of time (with zero initial conditions). We apply a constant torque $T = 0.25$ N m to the system. We will carry out an experiment both in vacuum and in a viscous medium, changing the damping coefficient.

Graphs of the angular displacement of the disk are shown in Fig. 3.31. Harmonic oscillation corresponds to modeling in a vacuum. By gradually increasing the viscosity, we achieve critical attenuation, i.e., rotation under the influence of external torque at a constant angle without hesitation.

Of particular interest is the phase diagram of the torsion oscillator. As mentioned above, the presence of constant exposure leads to the replacement of variables in the mathematical model, i.e., shift along the horizontal axis by a constant value. In this case, this is the angle of rotation of the disk, which we found numerically from Fig. 3.31. The phase diagram of a harmonic oscillator without friction is an external displaced ellipse, illustrated by displaced coils of a spiral, as shown in Fig. 3.32.

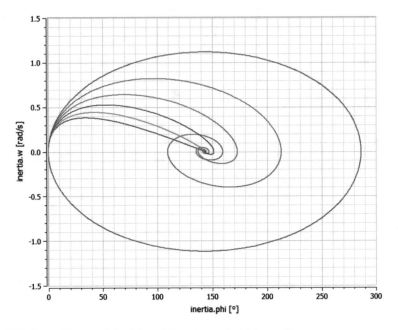

Fig. 3.32 Phase diagram of the disk at different viscosity of the medium

References

1. N.N. Bautin, E.A. Leontovich, Metody i priyomy kachestvennogo issledovaniya dinamicheskih sistem na ploskosti. – Moscow, Nauka (1990)
2. I.N. Efimov, E.A. Morozov, K.M. Selivanov, Computer simulation of dynamic systems. – Raleigh, North Carolina, USA: Lulu Press (2015)
3. Yu.B. Kolesov, Yu.B. Senichenkov, Matematicheskoe modelirovanie gibridnyh dinamicheskih system. – St. Petersburg, Izdatelstvo SPbPU (2014)
4. Yu.B. Kolesov, Yu.B. Senichenkov, Modelirovanie sistem. Dinamicheskie i gibridnye sistemy. – St. Petersburg, BHV-Peterburg (2012)
5. Yu.B. Kolesov, Yu.B. Senichenkov, Modelirovanie sistem. Ob'ektno-orientirovannyj podhod. – St. Petersburg, BHV-Peterburg (2012)
6. Yu.B. Kolesov, Yu.B. Senichenkov, Matematicheskoe modelirovanie slozhnyx dinamicheskix sistem. – St. Petersburg, Izdatelstvo SPbPU (2018)
7. Yu.V. Shornikov, D.N. Dostovalov, Osnovy modelirovaniya nepreryvno-sobytijnyh sobytij. – Novosibirsk, Izdatelstvo NGTU (2018)
8. A.A. Samarskii, A.P. Mikhailov, Principles of Mathematical modeling: Ideas, Methods, Examples. – CRC Press, Taylor & Francis Group (2002)

Chapter 4
Modeling of Mechanical Oscillatory Systems with One Degree of Freedom

Oscillatory movement, or simply oscillations, is called any movement or change of state, characterized by a varying degree of repeatability over time of the values of the physical quantities that determine this movement or state. We encounter oscillations in the study of various physical phenomena: sound, light, alternating currents, radio waves, pendulum swings, etc. It turns out that there is a commonality of the laws of these phenomena and mathematical methods for their study. Examples of oscillatory motion are oscillations of pendulums, strings, telephone membranes, charge, and current in an oscillatory circuit, etc.

More details about oscillatory systems can be found, for example, in references [1–3].

Oscillations are accompanied by the alternate conversion of the energy of one type into the energy of another type. Oscillatory motion is called periodic if the values of physical quantities that change during oscillations are repeated at regular intervals. All types of vibrations can be classified by the following parameters:

- by physical nature (mechanical and electromagnetic);
- by the nature of occurrence and existence (free, forced, parametric, and self-oscillations);
- by the nature of the dependence of the oscillating quantity on time (harmonic and non-harmonic).

Despite the different nature of the oscillations, the same physical laws are found in them; they are described by the same equations, investigated by general methods. In this section, we consider mechanical vibrations, i.e., repeated changes in the positions and velocities of any bodies or parts of bodies that occur in the presence of elastic forces, gravity, and other forces.

In the two previous chapters, the principles of constructing basic oscillatory models—the mathematical and spring pendulum—were considered. Now, we will move on to a more comprehensive study of their dynamic characteristics and the construction of more complex oscillatory models on their basis.

Let us start with the study of mathematical pendulum models.

© Springer Nature Singapore Pte Ltd. 2020
K. Rozhdestvensky et al., *Computer Modeling and Simulation of Dynamic Systems Using Wolfram SystemModeler*,
https://doi.org/10.1007/978-981-15-2803-3_4

4.1 Mathematical Pendulum

Formulation of the problem

Let a pendulum load of mass $m = 5$ kg, located at the end of a rod of length $l = 10$ m, be suspended on a fixed hinge. The hinge is considered perfectly smooth, the rod is considered weightless and absolutely rigid. The load is small in comparison with the length of the rod (material point), the acceleration of gravity g is constant, we will neglect air resistance at first, the vibrations occur in a fixed vertical plane. The load was rejected at an angle $\theta = 2°$ and released (Fig. 4.1).

Tasks

- In the Wolfram SystemModeler package, build nonlinear and linear (approximate) models of such a system in the absence of environmental resistance;
- Conduct numerical experiments with the constructed models, finding out at what angles of deviation the approximate linear model stops working;
- Repeat the experiment in a viscous medium. Choose the model parameters so that at least 12–15 periods of damped oscillations are observed. Build a graph of the amplitude versus time;
- Find out at what value of the attenuation coefficient the process ceases to be periodic (a pendulum, taken out of equilibrium, simply returns to it without oscillations—an aperiodic process).

Modeling and computational experiment

Since our model is described by a single equation in which nothing changes (no events occur), it is logical to attribute it to isolated continuous models.

Neglecting the resistance, we describe the motion of the pendulum by the nonlinear equation derived in the third chapter:

Fig. 4.1 Mathematical pendulum

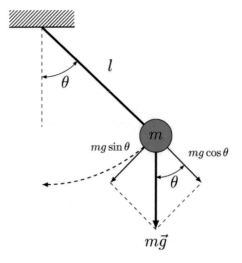

$$\frac{d^2\theta}{dt^2} + \frac{g}{l}\sin\theta = 0$$

where $\theta = \theta(t)$ is the angle of deviation of the pendulum from the vertical. We rewrite the equation as a system of two equations and express the derivatives:

$$\frac{d\theta}{dt} = \omega$$
$$\frac{d\omega}{st} = -\frac{g}{l}\sin\theta$$

The new variable ω will characterize the angular velocity.

Compare the simulation results with a linear model, which is described by the equation:

$$\frac{d^2\theta_1}{dt^2} + \frac{g}{l}\theta_1 = 0,$$

Let us build two models in text mode in Wolfram SystemModeler. Declare variables and constants. You can immediately set initial values for variables using the *start* attribute after the variable name. In both models, we will set the same initial values. Then, we write the equations. They are written in the *equation* block. The time derivative is specified using the *der()* function. The whole process was described in detail in the previous chapter, using the example of a spring pendulum.

In addition to obtaining the basic graphical dependencies, in the study of dynamic processes, we want to observe the movement of an animated 3D model. When working directly with code without using ready-made components, you will not see the animation. However, the movement of any of your dynamic models can be visually represented using the "*Visualizers*" component. Let us see how this is done.

First, as usual, we set constants, parameters, and variables.

```
constant Real pi = 3.1416 "Pi";
constant Real g(unit = "m/s2") = 9.81 "gravitational acceleration";
parameter Real l(unit = "m") = 10 "length of the thread";
parameter Real rad(unit = "m") = 0.5 "radius of the bob";
Real theta(unit = "rad", start = -pi / 90) "angular displacement";
Real omega(unit = "rad/s", start = 0) "angular velocity";
```

For visualization, in addition to the angular variables θ and ω, we will also need the Cartesian variables $x(\theta, \omega)$ and $y(\theta, \omega)$.

```
Real x(unit = "m") "horizontal displacement of the bob";
Real y(unit = "m") "vertical displacement of the bob";
```

The equations of the nonlinear model will obviously be as follows:

equation
  der(theta) = omega;
  der(omega) = -g / l * sin(theta);
  x = l * sin(theta);
  y = -l * cos(theta);

Now, you need to add the visualization of the pendulum. To do this, select in the Modelica library a section of ready-made components of the *Mechanics*, then *Multibody*, and finally *Visualizers*. Next, select *Advanced* and then *Shape* (Fig. 4.2.).

By double-clicking on the Shape button, we get a list of parameters that must be set in the program (Fig. 4.3.).

So the form search link looks like **Modelica.Mechanics.MultiBody.Visualizers. Advanced.Shape**. Since we do not use component modeling, but work in text mode, we will use Table 4.1 as reference information to create the corresponding lines of

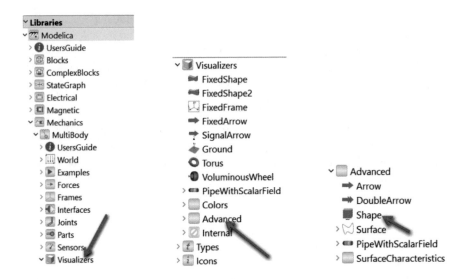

Fig. 4.2 Choosing a form for model visualization

Fig. 4.3 List of parameters for the Visualizer Shape

Table 4.1 Specifications for the shape, geometry, and color of the 3D model

Form name	Shape type	Geometric characteristics	Color (RGB)
Bob	Sphere	Length = 2 * rad, Width = 2 * rad, Height = 2 * rad, r = {x, y, 0}, r_shape = {-rad, 0, 0}	Color = {0, 50, 255}
Thread	Cylinder	Length = 1, Width = 0.2, Height = 0.2, LengthDirection = {1 * sin(theta), −1 * cos(theta), 0}	Color = {40, 100, 100}
Level	Box	Length = 1, Width = 0.1, Height = 0.1, r = {−1/2, 0, 0}	Color = {20, 20, 20}

code. We will describe three forms based on the information specified in the table.
Three additional lines should appear in the program:

```
Modelica.Mechanics.MultiBody.Visualizers.Advanced.
Shape bob(shapeType = "sphere", length = 2 * rad,
width = 2 * rad, height = 2 * rad, r = {x, y, 0}, r_shape =
{-rad, 0, 0}, color = {0, 50, 255});
Modelica.Mechanics.MultiBody.Visualizers.Advanced.
Shape thread1(shapeType = "cylinder", length = 1,
width = 0.2, height = 0.2, lengthDirection =
{1 * sin(theta), -1 * cos(theta), 0}, color = {40, 100, 100});
Modelica.Mechanics.MultiBody.Visualizers.Advanced.
Shape level(shapeType = "box", length = 1, width =
0.1, height = 0.1, r = {-1 / 2, 0, 0}, color = {20, 20, 20});
```

We add to the exact model of the mathematical pendulum we constructed an
approximate model for small angles. To do this, we introduce the variables θ_1, ω_1,
and the variables $x_1(\theta_1, \omega_1)$ and $y_1(\theta_1, \omega_1)$. For visualization, add a second rod and
a second mass

We carry out a computational experiment with the initial data specified in the
condition of the problem. The deviation angle $\theta = 2°$ can obviously be considered
small, so we expect a coincidence in the behavior of the two models. By smoothly
changing the initial displacement angles, we pass to the limiting case of large angles
and verify the divergence of the graphs. Figure 4.4 shows these two extreme cases.
The reader is invited to perform a full series of experiments on their own, paying
attention to the value of the initial angular displacement, at which the discrepancy in
the graphs becomes noticeable.

Run the animation of the model. You should see the movements of the two pendulums as in Fig. 4.5. When animating, the discrepancy in the behavior of the exact
and approximate models becomes even more obvious.

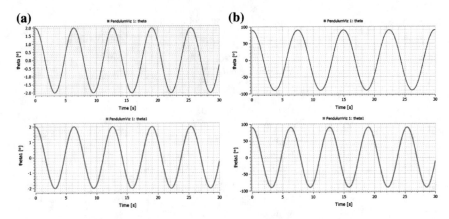

Fig. 4.4 Graph **a** shows the angular displacements in the exact (blue) and approximate (red) models with an initial deviation of 2°, **b** the angular displacements with a deviation of 90°—the oscillation period in the exact model increased, while in the approximate it remained unchanged

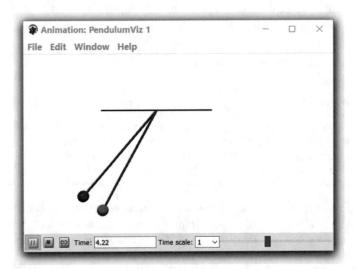

Fig. 4.5 Discrepancy between the exact (blue) and approximate (red) models

We turn to the modeling of damped oscillations. In Fig. 4.6, an experimental search for the "critical" attenuation coefficient is shown, in which the system returns to the equilibrium position without oscillations. The first experiment was conducted with a viscosity coefficient of 0.5 Ns/m (blue graph), the oscillations stopped at 1.5 Ns/m (green graph).

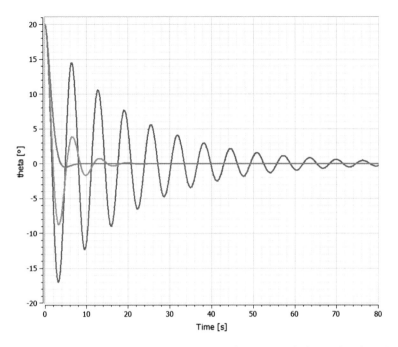

Fig. 4.6 Series of graphs of the damped oscillations of a mathematical pendulum in a viscous medium with a viscosity coefficient of 0.5, 1.0, and 1.5 Ns/m

Addition: complete code for accurate and approximate models

```
model PendulumViz
  constant Real pi = 3.1416 "Pi";
  constant Real g(unit = "m/s2") = 9.81
    "gravitational acceleration";
  parameter Real l(unit = "m") = 10
    "length of the thread";
  parameter Real rad(unit = "m") = 0.5 "radius of the bob";
  Real theta(unit = "rad", start = pi / 3)
    "angular displacement";
  Real omega(unit = "rad/s", start = 0) "angular velocity";
  Real x(unit = "m") "horizontal displacement of the bob";
  Real y(unit = "m") "vertical displacement of the bob";
  Real theta1(unit = "rad", start = pi / 3)
    "angular displacement1";
  Real omega1(unit = "rad/s", start = 0) "angular velocity1";
  Real x1(unit = "m") "horizontal displacement of the bob1";
  Real y1(unit = "m") "vertical displacement of the bob1";
  /*visualization objects*/
  Modelica.Mechanics.MultiBody.Visualizers.Advanced.Shape
  bob(shapeType = "sphere", length = 2 * rad, width =
2 * rad, height = 2 * rad, r = {x, y, 0}, r_shape =
{-rad, 0, 0}, color = {0, 50, 255});
  Modelica.Mechanics.MultiBody.Visualizers.Advanced.Shape
```

```
   thread(shapeType = "cylinder", length = 1, width =
 0.2, height = 0.2, lengthDirection = {1 * sin(theta),
  -1 * cos(theta), 0}, color = {40, 100, 100});
  Modelica.Mechanics.MultiBody.Visualizers.Advanced.Shape
  bob1(shapeType = "sphere", length = 2 * rad, width =
 2 * rad, height = 2 * rad, r = {x1, y1, 0}, r_shape =
 {-rad, 0, 0}, color = {255, 0, 0});
  Modelica.Mechanics.MultiBody.Visualizers.Advanced.Shape
  thread1(shapeType = "cylinder", length = 1, width =
 0.2, height = 0.2, lengthDirection = {1 * sin(theta1),
 -1 * cos(theta1), 0}, color = {40, 100, 100});
  Modelica.Mechanics.MultiBody.Visualizers.Advanced.Shape
  level(shapeType = "box", length = 1, width = 0.1, height =
 0.1, r = {-1 / 2, 0, 0}, color = {20, 20, 20});
 equation
  der(theta) = omega;
  der(omega) = -g / 1 * sin(theta);
  x = 1 * sin(theta);
  y = -1 * cos(theta);
  der(theta1) = omega1;
  der(omega1) = -g / 1 * theta1;
  x1 = 1 * sin(theta1);
  y1 = -1 * cos(theta1);
   annotation(Diagram(coordinateSystem(extent =
 {{-148.5, -105}, {148.5, 105}}, preserveAspectRatio =
 true, initialScale = 0.1, grid = {5, 5})));
 end PendulumViz;
```

4.2   Galileo Pendulum

Formulation of the problem

The Galileo pendulum is a mathematical pendulum of length L, oscillating near a vertical wall into which a nail is driven in at a distance l below the suspension point Fig. 4.7.

Tasks

• Explain why this dynamic model is hybrid;
• Construct differential equations describing this system and find the boundary conditions at the point of discrete change of parameters;
• Build graphs of the dependence of displacement on the equilibrium position on time;
• Build a phase portrait of this oscillator.

Modeling and computational experiment

The fundamental difference between this model and the usual mathematical pendulum is the introduction of additional conditions in the equations. This is done using the means of the Modelica language. Details on the introduction of discrete conditions were described in the introduction on the example of a bouncing ball. First, we

Fig. 4.7 Galileo pendulum

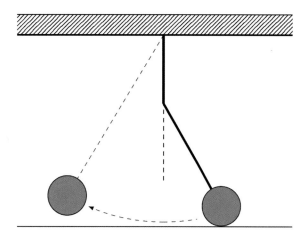

Fig. 4.8 State diagram of
the Galileo pendulum model

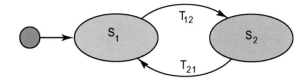

write the physical formulation of the problem: we divide the motion of the Galilean
mathematical pendulum into two stages.

We describe the mathematical model of this system in the form of a state diagram
(Fig. 4.8).

The **first stage** is the movement of the pendulum to the left of the equilibrium
position $\theta < 0$. We write down the equation of motion of a mathematical pendulum
in the field of gravity already known to us. We know that at this stage the pendulum
oscillates at full suspension:

$$\frac{d^2\theta}{dt^2} = -\frac{g}{L}\sin\theta$$

or, reducing to first-order equations:

$$\frac{d\theta}{dt} = \omega$$
$$\frac{d\omega}{dt} = -\frac{g}{l}\sin\theta$$

In software implementation:

```
der(theta) = omega;
if theta < 0 then
    der(omega) = -g / l * sin(theta);
```

```
x = 1 * sin(theta);
y = -1 * cos(theta);
```

The **second stage** is the movement of the pendulum to the right of the equilibrium position $\theta > 0$. The suspension length instantly changed and became equal to $L–l$. An event has occurred. New equation of motion:

$$\frac{d^2\theta}{dt^2} = -\frac{g}{L-l}\sin\theta$$

or

$$\frac{d\theta}{dt} = \omega$$

$$\frac{d\omega}{dt} = -\frac{g}{L-l}\sin\theta$$

In software implementation:

```
else
    der(omega) = -g / (1 - r) * sin(theta);
    x = (1 - r) * sin(theta);
    y = (-r) - (1 - r) * cos(theta);
end if;
```

At the time of the event, the linear velocity does not change, i.e., when moving to the right ($\theta < 0$), the discrete condition must be satisfied

$$L\omega = (L - l)\omega_1$$

What gives in software implementation redefinition of angular velocity:

```
when theta < 0 then
    reinit(omega, omega * (1 - r) / 1);
end when;
```

When moving to the left ($\theta > 0$), the discrete condition must be satisfied:

$$(L - l)\omega = L\omega_1$$

What gives in software implementation redefinition of angular velocity:

```
when theta > 0 then
    reinit(omega, omega * 1 / (1 - r));
end when;
```

Display the graphs of the main dependencies. On the graphs of displacement and speed, the moments of the onset of discrete events are clearly visible—they represent a "stitching" of sinusoids with different periods (Fig 4.9).

We will also construct a series of phase diagrams for the Galileo pendulum at various initial angles of deviation from the equilibrium position. They are "stitched" ellipses with different meanings (Fig. 4.10).

Addition: *full Galileo pendulum model code (without visualization lines)*

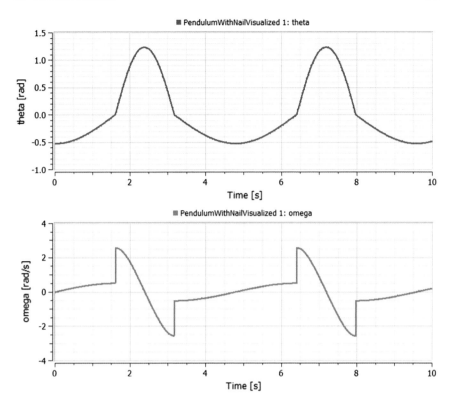

Fig. 4.9 Graphs of the deviation angle (blue) and angular velocity (red) versus time

```
model PendulumWithNailVisualized
  constant Real pi = 3.1416 "Pi";
  constant Real g(unit = "m/s2") = 9.81
    "gravitational acceleration";
  parameter Real l(unit = "m") = 10 "length of the thread";
  parameter Real r(unit = "m") = 8
    "distance from the pivot to the nail";
  parameter Real rad(unit = "m") = 0.5 "radius of the bob";
  Real theta(unit = "rad", start = -pi / 6)
    "angular displacement";
  Real omega(unit = "rad/s", start = 0) "angular velocity";
  Real x(unit = "m") "horizontal displacement of the bob";
  Real y(unit = "m") "vertical displacement of the bob";
equation
  der(theta) = omega;
  when theta < 0 then
           reinit(omega, omega * (l - r) / l);
  end when;
  when theta > 0 then
           reinit(omega, omega * l / (l - r));
  end when;
  if theta < 0 then
```

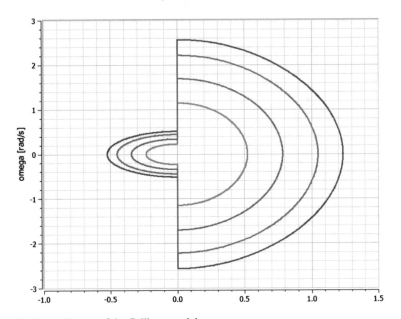

Fig. 4.10 Phase diagram of the Galilean pendulum

```
      der(omega)  =  -g / l * sin(theta);
   x = l * sin(theta);
   y = -l * cos(theta);
else
      der(omega)  =  -g / (l - r)  * sin(theta);
   x = (l - r)  *  sin(theta);
   y = (-r)  -  (l - r)  *  cos(theta);
   end if;
 end PendulumWithNailVisualized;
```

4.3 Mathematical Pendulum with Spring

Formulation of the problem

   On a weightless rod of length *l,* there is a load of mass *m*. A weightless spring of rigidity *k* is attached to an arbitrary point of the rod at a distance *d* from the suspension point (special cases are fastening in the middle of the rod or below the load). The second end of the spring is mounted on the wall (see Fig. 4.11). When the rod is upright, the spring is not deformed. Provide the ability to account for environmental resistance.

   Tasks

- Simulate the oscillations of a mathematical pendulum with a spring without taking into account friction. Build a graph of the change in angular displacement over

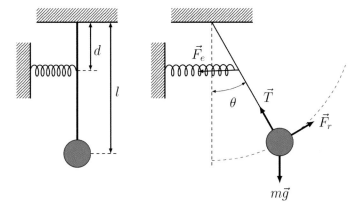

Fig. 4.11 Mathematical pendulum with a spring

time and compare it with the analytical and numerical solution of the oscillation problem of a conventional mathematical pendulum (use the results of Problem 4.1);

- Repeat the experiment by changing the position of the spring attachment point. As the main case, it is recommended to consider the fastening in the middle of the rod, as an additional—below the load;
- Compare the oscillation frequencies of the pendulums. Explain the result (here it is convenient to apply the law of conservation of mechanical energy in the absence of friction);
- Add resistance to the medium. Build graphs of damped oscillations. Is a case of critical attenuation possible here?

Modeling and computational experiment
When the pendulum deviates, a rotational moment arises, tending to return it to its equilibrium position. This moment is created by three forces: elasticity, gravity, and resistance of the medium. As a result, oscillations can occur at certain system parameters.

We write the equation of the dynamics of rotational motion for such a pendulum (for deviations smaller than the deviations leading to a revolution), taking into account all three points:

$$ml^2 \cdot \frac{d^2\theta}{dt^2} = -mgl \cdot \sin\theta - kd^2 \cdot \sin\theta \cdot \cos\theta - \gamma \frac{d\theta}{dt} \cdot l^2$$

Or, reducing to a system of equations of the first order, we obtain

$$\frac{d\theta}{dt} = \omega$$

$$\frac{d\omega}{dt} = \frac{-g}{l}\sin\theta - \frac{kd^2}{ml^2} \cdot \sin\theta \cdot \cos\theta - \frac{\gamma}{m}\omega$$

Table 4.2 Configurations of the system "Pendulum with a spring"

Experiment 1. The average position of the spring	Experiment 2. The highest upper position of the spring	Experiment 3. The lower position of the spring

We will carry out a series of computational experiments for different system configurations. The spring positions are shown on the 3D models of the spring pendulum made in WSM in Table 4.2.

When creating these experiments, it is convenient to save them under your own names (in our case, these will be PendulumWithSpring 1, PendulumWithSpring 2, PendulumWithSpring 3). Then, when changing the conditions of the experiment, for example, when taking into account the viscosity of the medium, it will be enough for you to call an experiment with the desired geometry of the model and make changes to the parameters.

For a clear comparison, in the numerical experiment, we impose the graphs of the displacement of the pendulum with the spring on the graphs of the usual mathematical pendulum, which we already obtained in Problem 4.1. It can be seen from the graphs that the limiting case of the upper arrangement of the spring coincides with a simple pendulum, in other cases, the influence of the spring leads to an increase in the oscillation frequency (verify this using an analytical calculation of the frequency and period of oscillation).

The results of numerical simulation are shown in Fig. 4.12.

Previous experiments were performed under the condition $\gamma = 0$, i.e., in a vacuum. Add the resistance of the medium and carry out the same experiments by clicking sequentially on the saved experiments PendulumWithSpring 1, PendulumWithSpring 2, PendulumWithSpring 3, and adding the viscosity value $\gamma = 0.5$ Ns/m.

The experimental results are shown in Fig. 4.13.

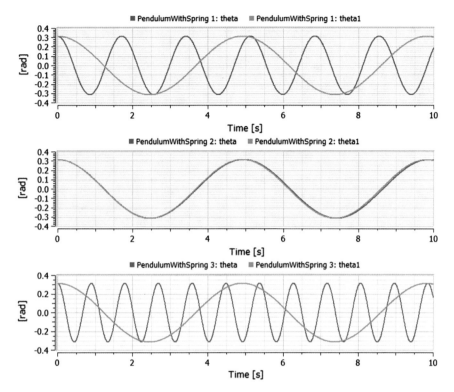

Fig. 4.12 Comparison of the angular displacements of a conventional mathematical pendulum and a pendulum with a spring at different positions of the spring attachment point. The upper graph corresponds to the average mounting position, the middle graph shows the coincidence of the oscillations with the oscillations of the mathematical pendulum at the extreme upper mounting position, and the lower graph shows the increase in the oscillation frequency with the lower mounting

Addition: full pendulum model code with spring

```
model PendulumWithSpring
  constant Real pi = 3.1416 "Pi";
  constant Real g(unit = "m/s2") = 9.81
    "gravitational acceleration";
  parameter Real l(unit = "m") = 6
    "length of the thread";
  parameter Real d(unit = "m") = 3
    "distance from the pivot to the spring";
  parameter Real m(unit = "kg") = 0.2 "mass";
  parameter Real k(unit = "N/m") = 10
    "coefficient of spring stiffness";
  parameter Real gamma(unit = "N s/m") = 0 "viscosity";
  parameter Real rad(unit = "m") = 0.5 "radius of the bob";
  Real theta(unit = "rad", start = pi / 10)
    "angular displacement";
  Real omega(unit = "rad/s", start = 0) "angular velocity";
```

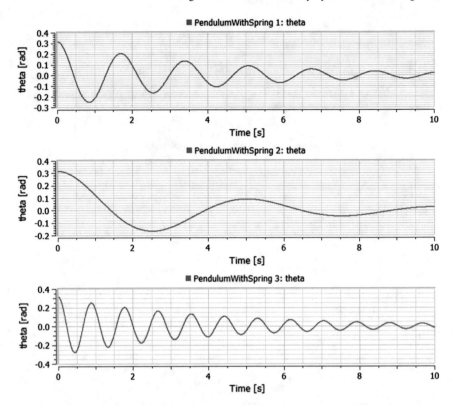

Fig. 4.13 Comparison of the angular displacements of the pendulum with the spring in a viscous medium, at different positions of the spring attachment point

```
Real x(unit = "m") "horizontal displacement of the bob";
Real y(unit = "m") "vertical displacement of the bob";
 /*visualization objects*/
Modelica.Mechanics.MultiBody.Visualizers.Advanced.
Shape bob(shapeType = "sphere", length = 2 * rad, width
= 2 * rad, height = 2 * rad, r = {x, y, 0}, r_shape =
{-rad, 0, 0}, color = {0, 50, 255});
  Modelica.Mechanics.MultiBody.Visualizers.Advanced.
Shape thread(shapeType = "cylinder", length = 1, width =
0.2, height = 0.2, lengthDirection = {x, y, 0}, color =
{40, 100, 100});
  Modelica.Mechanics.MultiBody.Visualizers.Advanced.
Shape level1(shapeType = "box", length = 1, width =
0.1, height = 0.1, r = {-1 / 2, 0, 0}, color = {20, 20, 20});
  Modelica.Mechanics.MultiBody.Visualizers.Advanced.
Shape level2(shapeType = "box", length = 0.1, width
= 3, height = 0.1, widthDirection = {0, 1, 0}, r =
{-1 / 2, -1 / 2, 0}, color = {20, 20, 20});
  Modelica.Mechanics.MultiBody.Visualizers.Advanced.
Shape spring(shapeType = "spring", extra = 10, r =
{-1 / 2, -1 / 2, 0}, length = 1 / 2+d * sin(theta) , width
```

```
= 0.2, height = 0.1, lengthDirection = {1, 0, 0}, color =
{20, 20, 20});
equation
  der(theta) = omega;
  der(omega) = -g / l * sin(theta)-k*d*d/(m*l*l)*sin(theta)
   *cos(theta)-gamma/m *omega;
  x = l * sin(theta);
  y = -l * cos(theta);
end PendulumWithSpring;
```

References

1. N.V. Butenin, Elementy teorii kolebanij. – Leningrad, Gosudarstvennoe soyuznoe izdatelstvo sudostroitelnoj promyshlennosti (1962)
2. N.V. Butenin, Elements of the Theory of Nonlinear Oscillations. – Blaisdell Publishing Company (1965)
3. A.A. Yablonskij, S.S. Norejko, Kurs teorii kolebanij. – Moscow, Gosudarstvennoe izdatelstvo «Vysshaya shkola» (1961)

Chapter 5
Modeling of Mechanical Oscillatory Systems with Several Degrees of Freedom

Until now, we have considered systems with one degree of freedom, having one natural frequency. If the design of the system is complicated, it can make a more complex movement. In many cases (in the absence of external forces), it can be reduced to the sum of two oscillations with different frequencies, depending on the properties of the system. Such a system has two degrees of freedom. Even a simple mathematical pendulum can oscillate in two mutually perpendicular directions, i.e., in the general case, it is a system with two degrees of freedom. Most often, several degrees of freedom are possessed by the so-called coupled systems—systems with many degrees of freedom—between which there are bonds that provide the possibility of energy exchange between different degrees of freedom. The main feature characteristic of any connected system is that its own oscillations are generally inharmonious and, depending on the method of observation, can be perceived either as beats that occur in such a way that the oscillation energy is periodically pumped (completely or partially) from one parts of the system to another and vice versa, or as the sum of two harmonic oscillations with frequencies ω^+ and ω^- determined by the structure of the system as a whole.

We begin the study of mechanical models with several degrees of freedom not from oscillation processes, but from the relative motion of two bodies. This is a comparative simple task that does not require any additional information, except for the basics of modeling studied in the previous chapter. The models given is this chapter are thoroughly considered in [1, 2].

5.1 The Movement of Two Bodies with Friction

The dynamic system consists of two bodies. The first body of mass m_1 moves along the horizontal plane, and the second body of mass m_2 moves along the first body with

© Springer Nature Singapore Pte Ltd. 2020
K. Rozhdestvensky et al., *Computer Modeling and Simulation of Dynamic Systems Using Wolfram SystemModeler*,
https://doi.org/10.1007/978-981-15-2803-3_5

Fig. 5.1 Movement of two
bodies with friction

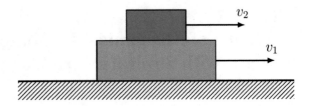

given initial velocities under the action of the friction forces that exist between all
the contacting bodies. The areas of application of the friction force are highlighted
in gray, as shown in Fig. 5.1.

Both movements are one-dimensional and have the same direction. It is assumed
that the friction force is proportional to the relative velocity of the two contacting
bodies.

Tasks

Build a model that will describe the change in the velocities of bodies, as well as
their position in space.

Conduct numerical experiments.

The initial velocity of the first body is $v_1(0) = 0.1$ m/s; the mass of the first body
is $m_1 = 100$ kg. The second body at the initial moment is at rest $v_2(0) = 0$; the mass
of the second body is $m_2 = 10$ kg.

For the same masses of bodies, the initial velocity of the second body is $v_2(0) =
0.2$ m/s. The first body at the initial moment is at rest $v_1(0) = 0$.

Modeling and computational experiment

We will consider the motion of bodies, as well as the change in velocities relative to
the horizontal surface; they will be equal to v_1 and v_2, respectively. We choose the
following conditional direction for speeds: We assume that speed is positive if the
body moves to the right, and negative if the body moves to the left.

The change in body speeds can be calculated using Newton's second law.

$$m\frac{dv}{dt} = F$$

We write these equations for each of the two bodies. We take into account that
friction forces with a horizontal surface and with a second body act on the first body.
The force acting on the second body is due only to friction with the first body.

$$\begin{cases} m_1\frac{dv_1}{dt} = -b_{s1} \cdot v_1 - b_{12} \cdot (v_1 - v_2) \\ m_2\frac{dv_2}{dt} = -b_{12} \cdot (v_2 - v_1) \end{cases}$$

$(v_1 - v_2)$ this is the speed of the first body relative to the second,

$(v_2 - v_1)$ this is the speed of the second body relative to the first.

The friction coefficients b_{s1} and b_{12} are the coefficients of friction with the surface
and the first body with the second, respectively.

```
TwoBody
model TwoBody
  parameter Real m1(unit = "kg") = 100 "body1";
  parameter Real m2(unit = "kg") = 10 "body2";
  parameter Real bs1(unit = "N s/m") = 18 "friction coefficient with surface";
  parameter Real b12(unit = "N s/m") = 16 "friction coefficient of bodies";
  Real x1(unit = "m", start = 0.0) "displacement1";
  Real v1(unit = "m/s", start = 0.1) "velocity1";
  Real x2(unit = "m", start = 0.0) "displacement2";
  Real v2(unit = "m/s", start = 0.0) "velocity2";
equation
  der(x1) = v1;
  m1 * der(v1) = - bs1 * v1-b12*(v1-v2);
  der(x2) = v2;
  m2 * der(v2) = -b12*(v2-v1);
  □;
end TwoBody;
```

Fig. 5.2 Program code for determining the change in the velocities of bodies, as well as their coordinates in space

For the coordinates of the bodies, we obviously have:

$$\begin{cases} \frac{dx_1}{dt} = v_1 \\ \frac{dx_2}{dt} = v_2 \end{cases}$$

So, if the position of the body at the initial moment of time is known, then their changes over time can be calculated using these formulas.

The software implementation is shown in Fig. 5.2.

We will carry out two computational experiments. In the first case, the first body first moves, and the second is at rest. The force of friction between the bodies makes the second body begin to move, increasing speed. When the relative velocity between the bodies becomes equal to zero, the friction between them disappears. However, the horizontal surface continues to slow down the first body and, consequently, decrease its speed. When the speed of the first body becomes less than the speed of the second body, the friction between the bodies begins to act again, but in this case, reducing the speed of the second body.

The graph of changes in the velocities of bodies is shown in Fig. 5.3.

In the second case, the first body is initially at rest and accelerates from the influence of the friction force from the side of the second body, as shown in Fig. 5.4.

5.2 Mechanical System with Damper and Spring

The coupled spring pendulums are an oscillating system composed of two spring pendulums (with masses m_1 and m_2, stiffnesses of springs k_1 and k_2, damping factors b_1 and b_2, respectively), fixed between two walls and interconnected by a spring with stiffness k. An external force (constant or periodic) can be applied to each mass, as shown in Fig. 5.5.

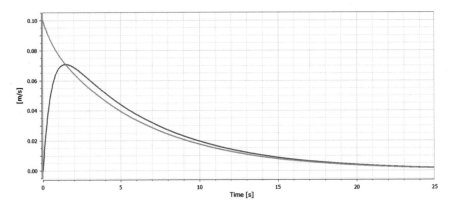

Fig. 5.3 Initial velocity of the first body is $v_1(0) = 0.1$ m/s; the mass of the first body is $m_1 = 100$ kg. The second body at the initial moment is at rest $v_2(0) = 0$; the mass of the second body is $m_2 = 10$ kg

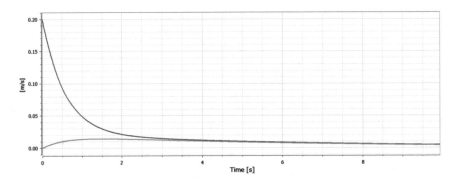

Fig. 5.4 Initial velocity of the second body is $v_1(0) = 0.2$ m/s. The first body at the initial moment is at rest $v_1(0) = 0$

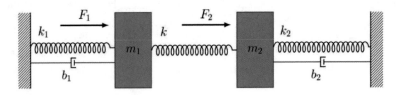

Fig. 5.5 Dynamic system: pendulums with spring and damper

Tasks

Build a mathematical model of this complex dynamic system. Explore its kinematic and dynamic characteristics.

Build graphs of the dependence of the position of the center of mass of both bodies on time.

Build graphs of the speed of bodies in a selected period of time, as well as graphs of changes in the deformation of each of the springs.

Conduct a series of experiments, changing the stiffness of the springs and the viscosity of the medium.

Calculate theoretically normal and partial oscillation frequencies of each of the bodies to simulate the conditions under which the oscillations of the bodies will be harmonic. In this case, it is possible to determine the oscillation frequency directly from the graph and compare them with the theoretical values obtained for normal frequencies. Validate results using FFT analysis.

Explore the resonance modes of a mechanical system.

Modeling and computational experiment

When modeling, it is assumed that the reader has read the previous sections on modeling the spring.

Let the distance between the walls be constant and equal to 1, the initial coordinate of the first body is determined by the length of the first spring in the undeformed state $x_{01} = l_1$. The initial position of the second body: $x_{02} = x_{01} + l_2$, where l_2 is the length of the second spring in the undeformed state. And finally, l_3 is the length of the third spring in an undeformed state

Using Newton's second law, we can write the equations of a free system as:

$$m_1 \frac{d^2 x_1}{dt^2} = -k_1(x_1 - x_{01}) - k((x_2 - x_1) - (x_{02} - x_{01})) - b_1 \frac{dx_1}{dt}$$

$$m_2 \frac{d^2 x_2}{dt^2} = -k((x_1 - x_2) - (x_{02} - x_{01})) - k_2(x_2 - x_{02}) - b_2 \frac{dx_2}{dt}$$

Here, the coordinates x_1 and x_2 are the positions of the centers of mass at an arbitrary point in time, so, for example, $\Delta x_1 = x_1 - x_{01}$ is the extension of the first spring. In addition, spring damping factors are taken into account here. If necessary, they can be set equal to zero. In addition to the differential equations in the mathematical model, the condition of fixed ends was used. Express the elongation of the springs through the displacement of the masses

$$\Delta x_1 = x_1 - x_{01}$$
$$\Delta x_2 = (x_2 - x_1) - (x_{02} - x_{01})$$
$$\Delta x_3 = x_2 - x_{02}$$

After deformation, the total length of the system should remain unchanged, i.e.,

$$(l_1 + \Delta x_1) + (l_2 + \Delta x_2) + (l_3 + \Delta x_3) = l$$

or

$$\Delta x_1 + \Delta x_2 + \Delta x_3 = 0$$

Add an external force to each equation and move on to first-order equations. Model of coupled spring pendulums:

$$\frac{dx_1}{dt} = v_1$$

$$m_1 \frac{dv_1}{dt} = -k_1(x_1 - x_{01}) - k((x_2 - x_1) - (x_{02} - x_{01})) - b_1 v_1 + F_1(t)$$

$$\frac{dx_2}{dt} = v_2$$

$$m_2 \frac{dv_2}{dt} = -k((x_1 - x_2) - (x_{02} - x_{01})) - k_2(x_2 - x_{02}) - b_2 v_2 + F_2(t)$$

When conducting computational experiments, the external force can be selected as follows:

– For free oscillations:

$$F_1(t) = F_2(t) = 0$$

– For forced oscillations with constant exposure:

$$F_1(t) = F_1$$

$$F_2(t) = F_2$$

– To study the phenomenon of resonance with external periodic exposure:

$$F_1(t) = F_1 \cos \omega_1 t$$

$$F_2(t) = F_2 \cos \omega_2 t$$

This mechanical system consists of components that are well known to us. It is on the example of these components in the second chapter that the features of software and component modeling were compared. Therefore, now we will no longer dwell on the technical subtleties, but give only the basic steps of analyzing the component model.

A diagram of a component model of a pendulum with two masses is shown in Fig. 5.6.

Animation of a computer model of a mechanical system consisting of two bodies, springs and dampers, can be seen during simulation of the model. An example of a screenshot of a system animation is shown in Fig. 5.7.

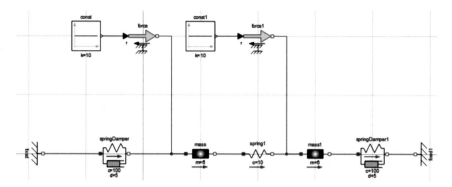

Fig. 5.6 Diagram of a component model of a pendulum with two masses

Fig. 5.7 Example
screenshot of a motion
animation of a mechanical
system

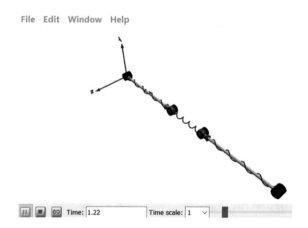

Without applying a driving force to the input, we carry out all the basic experiments—we study the behavior of each of the masses in a vacuum, achieving an
observation of the beats, and then we take into account the influence of viscosity and
construct graphs of damped oscillations.

We begin by studying the behavior of the system in a vacuum. Shown in Fig. 5.8
are graphs of the displacement of the first and second masses.

If we take the masses of springs differing 100 times, the nature of the oscillations
will noticeably change; see Fig. 5.9.

We take into account the viscosity of the medium and construct graphs of the
dependence of the position of the center of mass of both bodies on time. An example
of such a graph is shown in Fig. 5.10.

We plot the velocity of bodies in a selected period of time, as well as the graphs of
changes in deformation of each of the three springs of the system. It is recommended
to build velocity graphs of the first and second masses when observing harmonic
and non-harmonic oscillations. An example of the graphs is shown in Fig. 5.11.

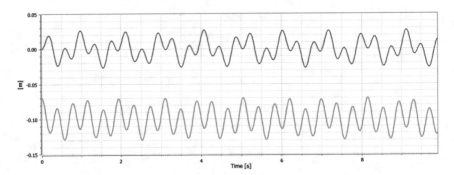

Fig. 5.8 Observation of beats: mass displacements at $m_1/m_2 = 1$ and opposite initial displacements of the springs $x_2 = -x_1$ in vacuum

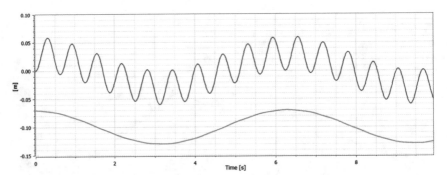

Fig. 5.9 Oscillations of different masses $m_1/m_2 = 100$ at nonzero initial displacement

Fig. 5.10 Graphs of the dependence of the position of the center of mass of both bodies on time (the upper graph is the left mass; the lower graph is the right mass)

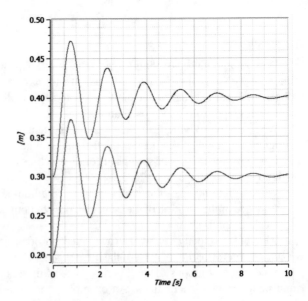

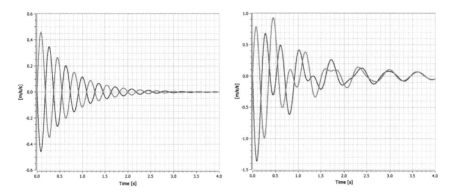

Fig. 5.11 Graphs of the velocities of the first (blue line) and second (yellow line) masses in harmonic and non-harmonic oscillations

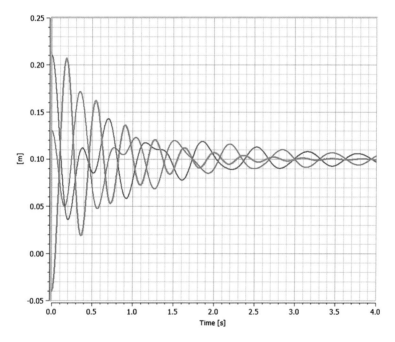

Fig. 5.12 Diagrams of deformation of three springs of a two-mass system during inharmonious oscillations—the first spring is a blue line, the second spring is green, and the third spring is orange

The mode of harmonic oscillations can be achieved by changing the value of the initial displacement of the springs.

The deformation graphs of the three springs of the system under study during non-harmonic oscillations are shown in Fig. 5.12.

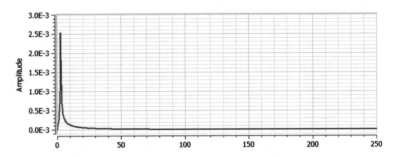

Fig. 5.13 Natural frequency of the first mass obtained using FFT analysis

Additionally, it is possible to plot the deformation of three springs of the investigated system under harmonic oscillations.

For this model, you can use the FFT analysis tool to determine the natural frequencies and compare with the results of analytical calculations and a numerical experiment. The eigenfrequency of the first mass obtained by FFT analysis is shown in Fig. 5.13.

In order to observe the resonance phenomenon in a two-mass system, it is necessary to replace both constant forces (Fig. 5.6) with periodic ones with a frequency that will coincide with the natural frequency of the system. In order to achieve resonance with the oscillations of the first mass, it is necessary to find the natural frequency of its oscillations according to the well-known formula:

$$\omega = \sqrt{\frac{k}{m}}$$

If you set the frequency of the external periodic force equal to the calculated angular frequency, then the external force will resonate with the first mass and spring. We calculate and establish such parameters for the system, taking into account that $\omega = 2\pi f$. The diagram of the component model for studying the resonance phenomenon is shown in Fig. 5.14.

To detect the resonance effect, it is necessary to increase the experiment time. You can do this by clicking the "Settings" tab and changing the stop time to 500 s. When choosing a numerical method, use the CVODES solver (Fig. 5.15).

5.3   The Movement of Three Bodies Connected by a Damper and Springs

Formulation of the problem

The system consists of two bodies with constant masses m_1 and m_2 moving on a horizontal surface, and a third body with constant mass m_3, which can move along

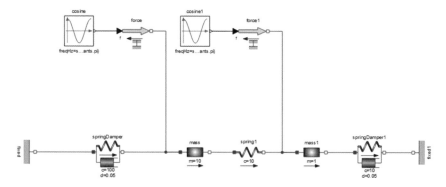

Fig. 5.14 Diagram of a component model for studying the resonance of a two-mass system

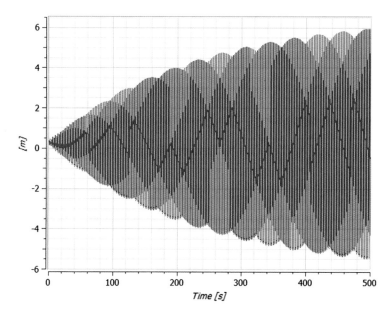

Fig. 5.15 Mass position charts

the first and second bodies. Between all contacting bodies, there is friction. A damper connects the first and second bodies, which are connected by springs to the walls. A predetermined external force F acts on the third body. The horizontal surface and side walls are motionless (Fig. 5.16).

Tasks

Build a model that will describe the change in the velocities of bodies, as well as their position in space.

Conduct numerical experiments:

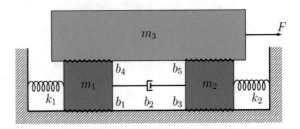

Fig. 5.16 Movement of three bodies connected by a damper and springs

In the first experiment, consider the value of the amplitude of the driving force equal to zero. In addition, there is no friction with the surface. The elastic coefficients of the springs are the same. To bring the system out of equilibrium, give the left body an initial speed of 5 m/s.

In the second experiment, add friction to the surface. Explain what you are observing.

Conduct an experiment with forcing periodic force. Build kinematic graphs to conclude.

Modeling and computational experiment

The first step in developing a model is to select a sign (direction) for speed. Let a body moving to the right have a positive speed and a body moving to the left should have a negative speed. We assume that the force is positive if it acts to the right.

Newton's second law allows you to connect the changes in the linear momentum of each body with the force applied to it.

$$\begin{cases} m_1 \frac{dv_1}{dt} = -b_1 \cdot v_1 - b_4 \cdot (v_1 - v_3) - b_2 \cdot (v_1 - v_2) - k_1 \cdot e_1 \\ m_2 \frac{dv_2}{dt} = -b_3 \cdot v_2 - b_5 \cdot (v_2 - v_3) - b_2 \cdot (v_2 - v_1) + k_2 \cdot e_2 \\ m_3 \frac{dv_3}{dt} = -b_4 \cdot (v_3 - v_1) - b_5 \cdot (v_3 - v_2) + F \end{cases}$$

Here e is the extension of the spring. The left end of the first spring is at rest, and the right end moves with speed v_1. The right end of the second spring is at rest, and the left end moves with speed v_2.

Consequently,

$$\begin{cases} \frac{de_1}{dt} = v_1 \\ \frac{de_2}{dt} = -v_2 \end{cases}$$

The relationship between the displacement of bodies and their speed can be expressed as follows:

$$\begin{cases} \frac{dx_1}{dt} = v_1 \\ \frac{dx_2}{dt} = v_2 \\ \frac{dx_3}{dt} = v_3 \end{cases}$$

The system model consists of eight equations; we consider the mass values of all bodies to be known (m_1, m_2, m_3), as well as the elastic coefficients of the springs (k_1, k_2). Damper viscosity coefficient (b_2), and friction coefficients (b_1, b_3, b_4, b_5).

The program code is shown in Fig. 5.17.

The position of the body at the initial time is selected as the reference point of the coordinate system for moving each body. Therefore, the initial positions of all three bodies are zero. Thus, the variable x of each body shows its displacement with respect to its initial position.

In the first experiment, we set the value of the amplitude of the driving force to zero. In addition, we assume that there is no friction with the surface. The elastic coefficients of the springs are the same. To bring the system out of equilibrium, it is necessary to give the left body an initial speed of 5 m/s. Oscillations will begin in the system; see Fig. 5.18.

The oscillations are close to in-phase and decay due to damping.

Now add friction to the surface. The bodies stop almost without oscillations, as shown in Figs. 5.19 and 5.20.

We will carry out an experiment with a forcing periodic force; see Fig. 5.21.

ThreeBodies

```
model ThreeBodies
  parameter Real m1(unit = "kg") = 100 "body1";
  parameter Real m2(unit = "kg") = 10 "body2";
  parameter Real m3(unit = "kg") = 15 "body3";
  parameter Real b1(unit = "N s/m") = 1.8 "friction coefficient 1";
  parameter Real b2(unit = "N s/m") = 3.2 "friction coefficient 2";
  parameter Real b3(unit = "N s/m") = 1.6 "friction coefficient 3";
  parameter Real b4(unit = "N s/m") = 1.8 "friction coefficient 4";
  parameter Real b5(unit = "N s/m") = 1.8 "friction coefficient 5";
  parameter Real k1(unit = "N/m") = 1.2 "coefficient of spring stiffness 1";
  parameter Real k2(unit = "N/m") = 1.2 "coefficient of spring stiffness 2";
  parameter Real F0(unit = "N") = 10 "amplitude of external force";
  parameter Real omega(unit = "Hz") = 0.1 "forcing frequency";
  Real x1(unit = "m", start = 0.0) "displacement1";
  Real v1(unit = "m/s", start = 0.0) "velocity1";
  Real x2(unit = "m", start = 0.0) "displacement2";
  Real v2(unit = "m/s", start = 0.0) "velocity2";
  Real x3(unit = "m", start = 0.0) "displacement3";
  Real v3(unit = "m/s", start = 0.0) "velocity3";
  Real F(unit = "N") "External force applied on body 3";
equation
  // Body 1
  der(x1) = v1;
  m1*der(v1) = -b1*v1 - b4*(v1 - v3) - b2*(v1 - v2) - k1*x1;
  // Body 2
  der(x2) = v2;
  m2*der(v2) = -b3*v2 - b5*(v2 - v3) - b2*(v2 - v1) - k2*x2;
  // Body 3
  der(x3) = v3;
  m3*der(v3) = -b4*(v3 - v1) - b5*(v3 - v2) + F;
  // External force
  F = F0*sin(omega*time);
  ¤;
end ThreeBodies;
```

Fig. 5.17 Movement of three bodies connected by a damper and springs

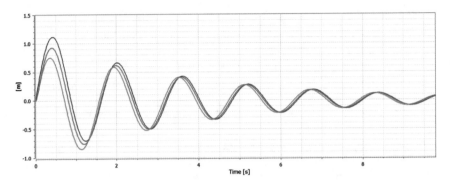

Fig. 5.18 An experiment with three bodies in the absence of driving force and friction with the surface. Three-body displacement graph

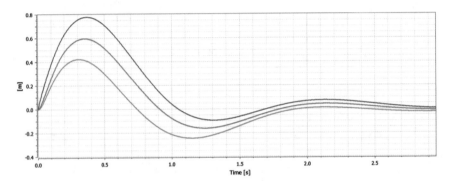

Fig. 5.19 An experiment with three bodies in the absence of driving force and in the presence of friction with the surface. Three-body displacement graph

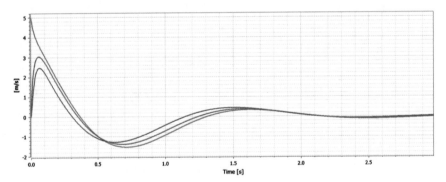

Fig. 5.20 An experiment with three bodies in the absence of driving force and in the presence of friction with the surface. Three-body velocity graph

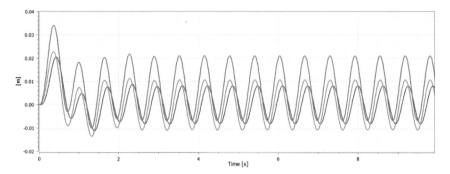

Fig. 5.21 An experiment with three bodies with constant exposure to periodic driving forces and in the presence of friction with the surface. Three-body displacement graph

In the graph shown in Fig. 5.21, the establishment of oscillations is clearly visible.

5.4 Connected Pendulums

Statement of the "asymmetric" problem

Two pendulums with masses m_1 and m_2 are fixed at points with different suspension heights with suspension lengths l_1 and l_2. Masses of pendulums are connected by a weightless spring with rigidity k. As an additional complication of the model, we can assume that different external forces F_1 and F_2 can act on each of the masses.

A diagram of the mechanical system under consideration is shown in Fig. 5.22. Two identical mathematical pendulums with masses m and with lengths of suspensions *l* are connected by a weightless spring with rigidity k. The spring is at a distance *d* from the suspension points located on one horizontal line.

The statement of the "symmetric" problem

Shown in Fig. 5.23 is a diagram of the mechanical system that is under consideration.

Tasks

Fig. 5.22 Dynamic system: two asymmetric pendulums, with masses connected by a spring

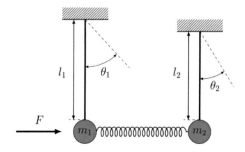

Fig. 5.23 Dynamic system:
two symmetrical pendulums,
with suspensions connected
by a spring

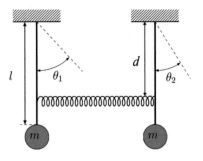

Select the spring stiffness coefficient, so that the coupling between the pendulums is weak by checking the corresponding condition. Calculate within the framework of the created model the normal oscillation frequencies and the beat frequency.

Construct graphs of the dependence of the angular displacement of each of the pendulums on one diagram. By changing the time interval and, if necessary, some of the parameters of the pendulums, achieve the appearance of a beat pattern (3–4 periods).

Investigate the influence of the stiffness coefficient of the coupling spring (at least 5–6 values) on the behavior of the system (period and beat frequency).

By changing the value of d (at least 5–6 values within $0 < d \leq l$), find out how the beat frequency changes.

Set nonzero initial deviations of the pendulums θ_{10}, $\theta_{20} \neq \theta$. Get oscillation patterns for several values of the initial displacements. Set according to the graph whether the amplitude of the displacements vanishes under such initial conditions? What does this mean from an energy point of view?

For the selected parameters of the pendulums, establish the initial deviations corresponding to harmonic oscillations. Experimentally (from the graphs) determine the period and calculate the circular frequency of the in-phase and antiphase oscillations. Compare the obtained values with normal oscillation frequencies.

Find out the effect of d on the frequencies of in-phase and antiphase oscillations (at least 8–10 values in the range $0 < d \leq l$). Compare experimental values (obtained directly from the oscillation graphs) with theoretical.

Break the system into two partial ones, analyze their behavior, and compose the equation of motion of any of them for the case of small deviations. Calculate the partial oscillation frequencies. Write down the solution of the resulting equation (the equation of harmonic oscillations with a frequency equal to the partial). Build a graph of the oscillations of the partial system by setting the initial data at which harmonic oscillations are observed. By changing the value of d, determine:

(1) in which case the pendulums oscillate with a frequency equal to the partial;
(2) what is the range of change of partial frequencies and compare it with the interval of normal frequencies $[\omega^{+}; \omega^{-}]$. Explain the results.

Investigate the behavior of the system with unequal parameters of the pendulums ($m_1 \neq m_2$ or $l_1 \neq l_2$). Check if harmonic oscillations are possible in this case.

Make and solve the equations of motion of the pendulums taking into account the force of resistance of the medium.

Modeling and computational experiment

When the pendulums move in a vertical plane, the state of the system is completely described by two independent parameters—the angles θ_1 and θ_2 of the deviations of the pendulums from the vertical. Thus, a mechanical system has two degrees of freedom. We obtain a system of equations, which can be described as a "symmetric" and an "asymmetric" system.

The equation of free motion for each pendulum is similar to the equation obtained above for a simple mathematical pendulum:

$$J\frac{d^2\theta}{dt^2} = M$$

In the absence of viscosity of the medium, the moment of rotation created by gravity acts on the first pendulum

$$M_{11} = -m_1 g l_1 \sin\theta_1$$

created by gravity, and the moment of elasticity

$$M_{12} = kd^2(\sin\theta_2 - \sin\theta_1)\cos\theta_1$$

Therefore, the equation of motion of the first pendulum will have the form:

$$m_1 l_1^2 \frac{d^2\theta_1}{dt^2} = -m_1 g l_1 \sin\theta_1 + kd^2(\sin\theta_2 - \sin\theta_1)\cos\theta_1$$

or

$$\frac{d^2\theta_1}{dt^2} = \frac{-g}{l_1}\sin\theta_1 + \frac{kd^2}{m_1 l_1^2}(\sin\theta_2 - \sin\theta_1)\cos\theta_1$$

When deriving the equation, it is assumed that the deflection angles of the pendulum are limited to 180°, i.e., considered the movement of the pendulum without a "coup." A similar formula is obtained for the second pendulum. Thus, for two pendulums connected by a spring in the absence of external forces, the system of equations:

$$\frac{d\theta_1}{dt} = \omega_1$$

$$\frac{d\omega_1}{dt} = \frac{-g}{l_1}\sin\theta_1 + \frac{kd^2}{m_1 l_1^2}(\sin\theta_2 - \sin\theta_1)\cos\theta_1$$

$$\frac{d\theta_2}{dt} = \omega_2$$

$$\frac{d\omega_2}{dt} = \frac{-g}{l_2} \sin\theta_2 - \frac{kd^2}{m_2 l_2^2}(\sin\theta_2 - \sin\theta_1)\cos\theta_2$$

where ω_1 and ω_2 are the corresponding angular velocities. In the particular case of small deviation angles, this system decomposes into two independent equations in which the variables are the sum of the angular displacements $\theta_1 + \theta_2$ and their difference $\theta_1 - \theta_2$. Moreover, each of these equations describes oscillations of a harmonic oscillator with natural frequencies ω^+ and ω^-, respectively.

When considering the "symmetric" system, it is necessary to put $l_1 = l_2, m_1 = m_2$ in these equations, when considering the "asymmetric" system for the first pendulum $l_1 = d$ and for the second pendulum $l_2 = d$.

We begin the simulation with a special case of the mathematical model constructed above for the case of a symmetric system, i.e., identical pendulums ($l_1 = l_2, m_1 = m_2$).

Figure 5.24 shows an example code for this model without visualization. This is due to the large volume of such code. However, a model with several degrees of freedom, as a rule, turns out to be quite complex and, to facilitate the interpretation of its behavior, it makes sense to perform visualization based on the examples described above.

We carry out the necessary numerical experiments.

Let us consider the behavior of the angles of displacement of the pendulums θ_1 and θ_2 for the case of weak coupling, when the action of the elastic force is noticeably less than the gravity $kd \ll mgl^2$. Consider the case when at first only one of the pendulums was rejected, i.e., at time $t = 0$, the displacement amplitude of the second pendulum θ_2 is zero

```
TwoPendulumwithSpring                                              |model    |Modelica Te:
model TwoPendulumwithSpring
    constant Real pi = 3.1416 "Pi";
    constant Real g(unit = "m/s2") = 9.81 "gravitational acceleration";
    parameter Real l1(unit = "m") = 1 "length of the thread1";
    parameter Real l2(unit = "m") = 1 "length of the thread2";
    parameter Real d(unit = "m") = 0.2 "displacement of the spring";
    parameter Real m1(unit = "kg") = 0.2 "mass1";
    parameter Real m2(unit = "kg") = 0.2 "mass2";
    parameter Real k(unit = "N/m") = 10 "coefficient of spring stiffness";
    Real theta1(unit = "rad", start = pi / 10) "angular displacement1";
    Real omega1(unit = "rad/s", start = 0) "angular velocity1";
    Real theta2(unit = "rad", start = 0) "angular displacement2";
    Real omega2(unit = "rad/s", start = 0) "angular velocity2";
equation
    der(theta1) = omega1;
    der(omega1) = -g / l1 * sin(theta1)+(k*d*d)/(m1*l1*l1)*(sin(theta2)-sin(theta1))*cos(theta1);
    der(theta2) = omega2;
    der(omega2) = -g / l2 * sin(theta2)-(k*d*d)/(m2*l2*l2)*(sin(theta2)-sin(theta1))*cos(theta2);
    n;
end TwoPendulumwithSpring;
```

Fig. 5.24 Program code of the model of coupled pendulums

We take $l_1 = l_2 = 1$ m, $m_1 = m_2 = 0.2$ kg. The value of the distance d at which we fix the spring, we take equal to 0.21. The initial values of the angles of deviation of the pendulums: $\theta_{10} \neq 0$ (within 5°–6°) and $\theta_{20} = 0$.

In this case, the first pendulum begins to oscillate alone, but over time, the amplitude of oscillations of the second pendulum θ_2 starts to increase at the moment when the amplitude θ_1 drops to zero, the amplitude θ_2 reaches its maximum value, etc. i.e., there is an exchange of energy.

This behavior is clearly visible in the graph, as shown in Fig. 5.25.

The same picture is observed for an arbitrary initial displacement of both pendulums. And only in two cases, there is no energy exchange between the pendulums:

(a) If two identical pendulums at the initial moment of time were deflected by the same angle ($\theta_1 = \theta_2$) and had the same speeds, then they perform in-phase harmonic oscillations with frequencies $\omega+$.

We will verify this using a numerical experiment. We set the initial conditions as indicated in Fig. 5.26.

The graph of Fig. 5.27 shows the in-phase angular displacement of two pendulums.

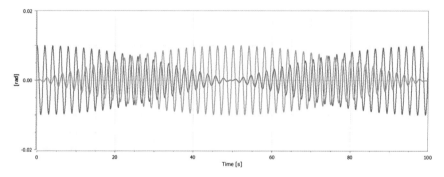

Fig. 5.25 Illustration of energy exchange between connected pendulums

Fig. 5.26 Selection of initial conditions for observing in-phase harmonic oscillations

Plot	Parameters	Variables	Settings

⊞ Filter

Name	Initial Val	Unit	Description
omega1	0.0	ra...	angular velocity1
omega2	0.0	ra...	angular velocity2
theta1	0.31416	rad	angular displacement1
theta2	0.31416 ˅ rad		angular displacement2

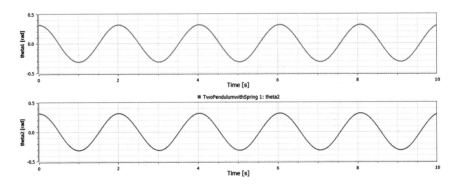

Fig. 5.27 In-phase harmonic oscillations of coupled pendulums

To study the nature of the oscillations in this case, it is convenient to display an additional visualization panel. Then, we can visually observe the in-phase oscillations of the pendulums.

(b) If at $t = 0$ the deviations and projections of the velocities are equal and opposite in sign, the pendulums will also perform harmonic oscillations with a frequency ω^+, but occurring in antiphase (antiphase oscillations).

Now we choose $\theta_1 = -\theta_2$ to observe the antiphase oscillations (Fig. 5.28).

To study the nature of the oscillations in this case, it is convenient to display an additional visualization panel. Then, we can visually observe the antiphase oscillations of the pendulums (Fig. 5.29).

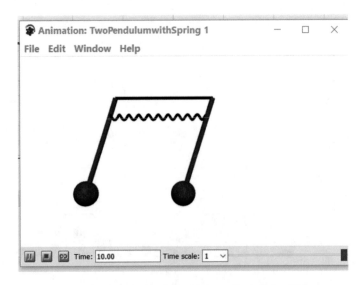

Fig. 5.28 3D animation of in-phase harmonic oscillations of coupled pendulums

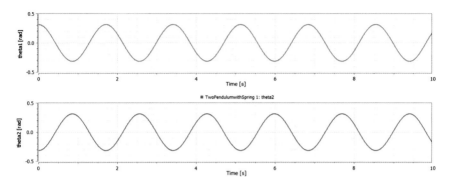

Fig. 5.29 Antiphase harmonic oscillations of coupled pendulums

These in-phase and antiphase oscillations are called normal modes of oscillations (or normal oscillations) of a system of coupled oscillators, and the frequencies ω^+ and ω^- are called normal frequencies. The normal mode of oscillations is the harmonic oscillations that each of the coupled pendulums performs with a special choice of initial conditions. With any other choice of the initial deviations in each of the pendulums, both normal oscillations appear at once; in other words, the arising oscillations are a superposition of two normal oscillations. This follows from the fact that any initial deviation of two pendulums can be represented as the sum of two initial deviations: one in which both pendulums are equally deflected in one direction, and the other, in which both pendulums are equally rejected in opposite directions. The weaker the spring connecting the pendulums, the obviously closer both normal frequencies will be to each other (Fig. 5.30).

The presence of weak coupling means that the frequency detuning $\Delta\omega = \omega^+ - \omega^-$ is small compared to the normal frequencies ω^+ and ω^-, and the periodic increase and decrease in the amplitude of oscillations of each of the pendulums (i.e., beating)

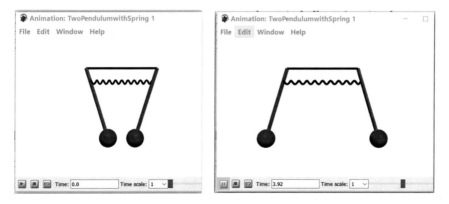

Fig. 5.30 3D animation of antiphase harmonic oscillations of coupled pendulums

occurring with the frequency $\Delta\omega$ (beat frequency) are easily observed experimentally. The oscillation amplitudes of each of the pendulums periodically change with a phase shift of $\pi/2$: When one of them reaches a maximum, the second vanishes and vice versa.

This behavior of the pendulums can be understood by appealing to the normal modes of oscillations. In the case of an even mode of normal oscillations, indicated by the "+" sign, the pendulums move together, the spring is not stretched, and the frequency is the same as for a single oscillator. In the case of an odd mode of normal oscillations (the "−" sign), the spring is stretched, which increases the frequency of this oscillation mode. If only one of the pendulums is displaced, we have two normal modes of oscillation, which are in a certain relative phase. But since the frequency of the odd oscillation is slightly higher than the frequency of the even oscillation, the relative phase changes.

After some time, two normal modes of oscillation will be in antiphase, the amplitude θ_1 will drop to zero, while the amplitude θ_2 will peak, etc.

You can also consider the same situation from an energy point of view. At $t = 0$, all energy is concentrated in the first pendulum. As a result of communication through the spring, energy is gradually transferred from the first pendulum to the second pendulum until all the energy has accumulated in the second pendulum. The frequency with which the oscillators exchange energy is equal to the difference in normal frequencies $\omega^+ - \omega^-$.

We should also mention the so-called partial (partial) oscillations inherent in coupled systems. These oscillations are obtained if one of the pendulums is rigidly fixed, and the second is removed from the equilibrium position and left to its own devices (without destroying the connection). Obviously, the partial frequency will exceed the frequency of one pendulum. In a symmetric system (identical pendulums), both partial frequencies are equal to each other, and this common value is enclosed between the values of two normal frequencies.

In the case of various pendulums, their motion is more complex than that described above for a symmetric system.

Now let the initial deviation and velocity values be different for different pendulums, as shown in Fig. 5.31.

In this case, the oscillations are not harmonic, as shown in Fig. 5.32.

We pass to the second part of the problem. We assemble an asymmetric system from ready-made components (Fig. 5.33).

Starting a numerical experiment in the Simulation Center window led to the creation of the following 3D animation (Fig. 5.34).

As mentioned above, in the asymmetric case, we will not be able to isolate normal oscillation modes; therefore, in this case we will conduct an experiment with a forcing periodic force with an amplitude of $F_x = 10$ N and a frequency of $f = 1$ Hz. As a result of the experiment, we obtain graphical dependences of the kinematic parameters of the motion of the coupled pendulum on time. A graph of the change in the angles of displacement of the pendulums as a function of time (20 s) is shown in Fig. 5.35.

Phase diagrams of both pendulums are shown in Fig. 5.36.

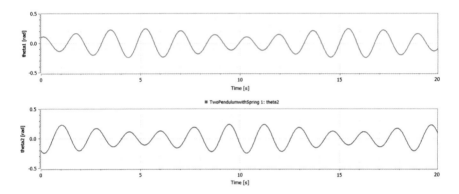

Fig. 5.31 Arbitrary initial values

Fig. 5.32 Observation of non-harmonic oscillations

As can be seen from the above figures, under these initial conditions, the oscillations of the pendulums are not harmonic, but they occur in antiphase. The oscillations of the pendulums are of a repeating nature, and the oscillations of the second pendulum are of greater scope. This is due to the action of external force on the first pendulum.

We remove the external force, and as the initial condition we give the first pendulum the initial angular velocity $\omega_{x0} = 2$ rad/s. The results are presented in Figs. 5.37 and 5.38 (simulation time—20 s)

Oscillations are still not harmonious. However, the phase diagrams show that the pendulums have two "centers" (or a segment along the abscissa axis) with respect to which oscillations occur, which roughly corresponds to the pattern of beats. Without the influence of an external force, the swing ranges of the pendulums became approximately the same.

It can be assumed that, if the swings of the oscillations of the pendulums are reduced (i.e., go into the region of "small oscillations"), then the oscillations themselves should approach harmonic ones in character. For this, we take the angular

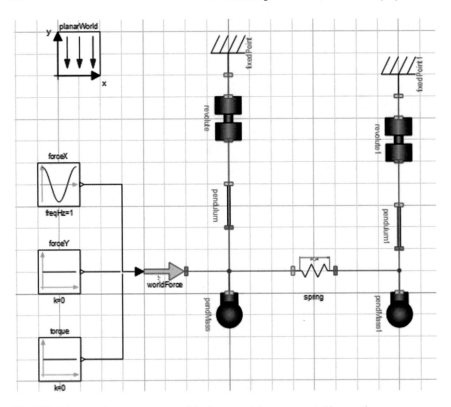

Fig. 5.33 Diagram of a component model of two pendulums connected by a spring

Fig. 5.34 3D animation of
the created computer model

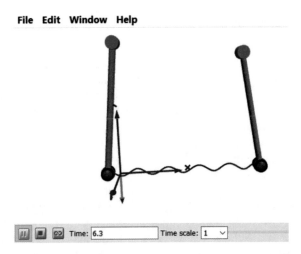

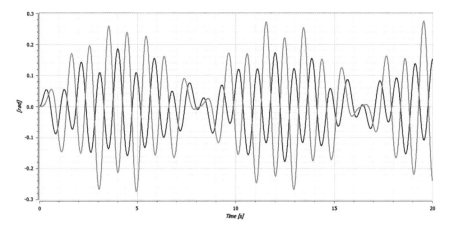

Fig. 5.35 Graphs of the dependence of the angles of displacement of mathematical pendulums on time: blue—the first pendulum, yellow—the second pendulum

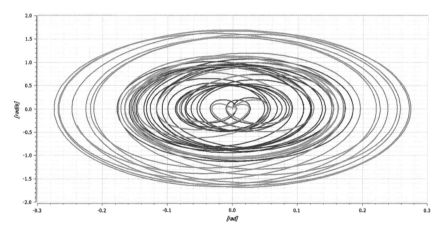

Fig. 5.36 Phase diagrams of mathematical pendulums: blue—the first pendulum, yellow—the second pendulum

velocity $\omega_{x0} = 0.2$ rad/s as the initial condition. The results are shown in Figs. 5.39 and 5.40 (simulation time—20 s).

As can be seen from the figures, under such an initial condition, the pendulums indeed do small oscillations, since the angles of displacement in absolute value do not exceed 0.06 rad ($= 2.9°$). At the same time, the nature of the phase diagrams did not change qualitatively; the oscillations remained inharmonic.

If you build a graph of the angle of displacement of one pendulum from the angle of displacement of another, you can get an interesting graph (Fig. 5.41). From this graph, the envelope of the phase trajectories of the motion of the coupled pendulum

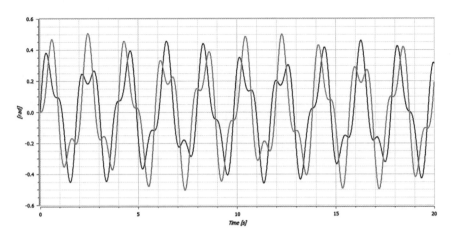

Fig. 5.37 Graphs of the dependence of the angles of displacement of the pendulums on time: blue—the first pendulum, yellow—the second pendulum ($\omega_{x0} = 2$ rad/s)

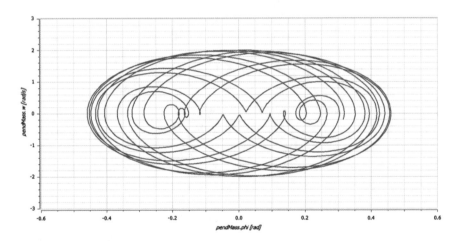

Fig. 5.38 Phase diagram of the first pendulum ($\omega_{x0} = 2$ rad/s)

in the plane of the displacement angles is clearly visible. It is a rhombus, offset from the origin.

Similar envelopes of phase trajectories can also be obtained in the plane of linear displacements of pendulums from the equilibrium position.

Let us return to the initial settings of the system parameters with the frequency of the driving force in the X-direction equal to $f = 1$ Hz. Theoretical calculations give us two values of natural frequencies:

$$\omega_1 = 0.511964 \, \text{Hz}$$
$$\omega_2 = 1.12947 \, \text{Hz}$$

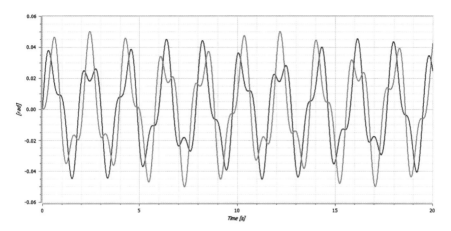

Fig. 5.39 Graphs of the dependence of the angles of displacement of the pendulums on time: black—the first pendulum, gray—the second pendulum ($\omega_{x0} = 0.2$ rad/s)

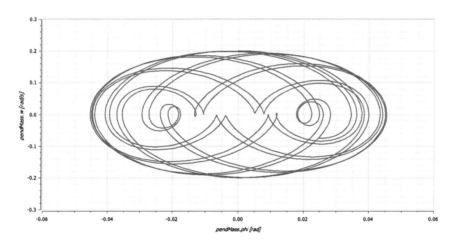

Fig. 5.40 Phase diagram of the first pendulum ($\omega_{x0} = 0.2$ rad/s)

For the experimental determination of the eigenfrequencies of each pendulum, one should use the Fourier transform to switch from the time scale to the frequency one. Figures 5.42 and 5.43 show the frequency response and phase response for the first and second pendulums. The graphs show that for the studied system of pendulums, interconnected by a spring, under the selected initial conditions, the characteristic frequencies are $f_1 = 0.5$ Hz and $f_3 = 1.15$ Hz.

In addition, a clear peak is observed at the frequency of the driving force $f = 1$ Hz.

Moreover, the analysis of the phase response shows that the oscillations of the pendulums are in antiphase.

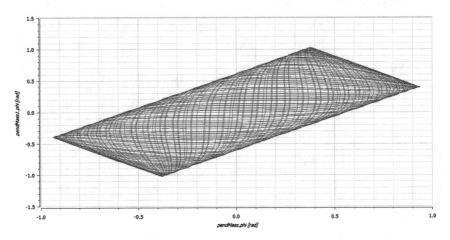

Fig. 5.41 Envelope phase diagram of coupled pendulums

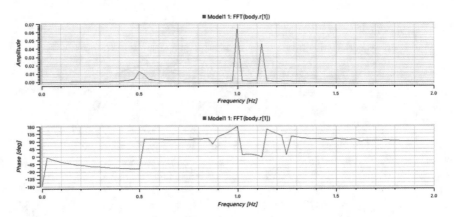

Fig. 5.42 Graphs of the amplitude–frequency and phase–frequency characteristics of the pendulum 1: above—frequency response, bottom—phase response

Thus, we can conclude that the calculated frequencies correspond to the frequencies that we see in the graphs obtained in the FFT analysis.

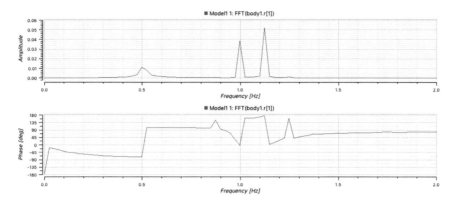

Fig. 5.43 Graphs of the amplitude–frequency and phase–frequency characteristics of the pendulum 2: above—frequency response, bottom—phase response

5.5 Double Pendulum

Formulation of the problem

A double pendulum can be represented as follows: A single mathematical pendulum at the point of load location m_1 has a swivel, to which a second mathematical pendulum with a load m_2 is suspended, on an inextensible thread of length l_2, forced to swing in the same plane, as shown in Fig. 5.44

Tasks

To simulate the oscillations of each body individually, i.e., calculate the velocities of bodies and their angular displacements (θ_1 and θ_2).

Build graphs of the dependence of the angles of deviation of bodies on time in one diagram.

Fig. 5.44 Dynamic system: double pendulum

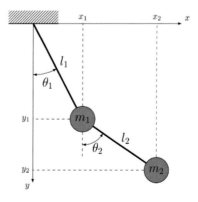

Modeling and computational experiment

Compilation of the differential equation of motion of a double pendulum in Cartesian coordinates is difficult due to the presence of reactions that occur in the hinge joints. Similar problems are solved by compiling equations of motion for generalized coordinates (using the Lagrange equations). The fact is that setting the position of a system of points fastened by bonds in Cartesian coordinates is not always convenient. The choice of parameters necessary to describe the position of all points of the mechanical system (i.e., generalized coordinates) should be determined primarily by expediency.

In our case, it is convenient to take the angles of deviation of each of the pendulums from the vertical (θ_1 and θ_2) as generalized coordinates.

First, it is necessary to introduce a reference frame and express the Cartesian coordinates in terms of generalized:

$$\begin{cases} x_1 = l_1 \sin \theta_1 \\ y_1 = -l_1 \cos \theta_1 \end{cases} \quad \text{and} \quad \begin{cases} x_2 = l_1 \sin \theta_1 + l_2 \sin \theta_2 \\ y_2 = -(l_1 \cos \theta_1 + l_2 \cos \theta_2) \end{cases}$$

Then, by differentiating these equalities, we obtain the Cartesian velocity components expressed in terms of the generalized coordinates (θ_1 and θ_2) and the generalized velocities ($\dot{\theta}_1$ and $\dot{\theta}_2$):

$$\begin{cases} \dot{x}_1 = l_1 \dot{\theta}_1 \cos \theta_1 \\ \dot{y}_1 = l_1 \dot{\theta}_1 \sin \theta_1 \end{cases} \quad \text{and} \quad \begin{cases} \dot{x}_2 = l_1 \dot{\theta}_1 \cos \theta_1 + l_2 \dot{\theta}_2 \cos \theta_2 \\ \dot{y}_2 = l_1 \dot{\theta}_1 \sin \theta_1 + l_2 \dot{\theta}_2 \sin \theta_2 \end{cases}$$

Then, you can calculate the squares of the speeds of each pendulum (included in the kinetic energy):

$$\begin{cases} v_1^2 = \dot{x}_1^2 + \dot{y}_1^2 \\ v_2^2 = \dot{x}_2^2 + \dot{y}_2^2 \end{cases}$$

Cartesian coordinates included in the potential energy formula,

$$W = m_1 g y_1 + m_2 g y_2$$

They are replaced with generalized coordinates. The velocities included in the kinetic energy formula should be replaced by generalized velocities, so that kinetic energy in the general case will depend on the generalized coordinates

$$T = \frac{m_1 v_1^2}{2} + \frac{m_2 v_2^2}{2}$$

Write down the Lagrangian, and compose the Lagrange equations for the first and second pendulums in the absence of non-potential forces:

$$\begin{cases} \frac{\partial}{\partial t}\frac{\partial L}{\partial \dot\theta_1} - \frac{\partial L}{\partial \theta_1} = 0 \\ \frac{\partial}{\partial t}\frac{\partial L}{\partial \dot\theta_2} - \frac{\partial L}{\partial \theta_2} = 0 \end{cases}$$

After algebraic transformations, it is easy to obtain the following system of differential equations for the mathematical model of a double pendulum.

$$\frac{d\theta_1}{d}t = \omega_1$$

$$\frac{d\omega_1}{dt} = -\frac{m_2}{m_1 + m_2 \sin^2(\theta_1 - \theta_2)}\left[\sin(\theta_1 - \theta_2)\left(\dot\theta_1^2 \cos(\theta_1 - \theta_2) + \frac{l_2}{l_1}\cdot\dot\theta_2^2\right)\right.$$
$$\left. - \frac{g}{l_1}\left(\sin\theta_2 \cos(\theta_1 - \theta_2) - \frac{m_1 + m_2}{m_2}\sin\theta_1\right)\right]$$

$$\frac{d\theta_2}{d}t = \omega_2$$

$$\frac{d\omega_2}{dt} = -\frac{m_2 + m_1}{m_1 + m_2 \sin^2(\theta_1 - \theta_2)}\left[\sin(\theta_1 - \theta_2)\left(\frac{m_2}{m_1 + m_2}\dot\theta_2^2 \cos(\theta_1 - \theta_2) + \frac{l_1}{l_2}\cdot\dot\theta_1^2\right)\right.$$
$$\left. + \frac{g}{l_2}(\sin\theta_1 \cos(\theta_1 - \theta_2) - \sin\theta_2)\right]$$

This system is written for the case of free oscillations. In the case of external non-potential forces, the corresponding term Q_{ext}^n must be added to Lagrangian.

We do not provide the program code here due to its large size. After the successive creation of previous models, the reader will easily write it, relying on the mathematical model of the problem.

We begin the experiment with small deflection angles. In this case, a process close to beating can be observed. The masses in this case move in antiphase. At large deviation angles and nonzero initial velocities, the nature of the oscillations changes (Fig. 5.45).

For a better understanding of the process, it makes sense to run a model animation with a path trace (Fig. 5.46).

We give another example of a phase diagram for the second mass (Fig. 5.47).

5.6 Dual Torsion Oscillator

Formulation of the problem

A double torsion spring oscillator consists of two parallel disks (or rotors) that can rotate relative to a fixed axis perpendicular to the planes in which the disks are located. The disks are coupled to a weightless coil spring (Fig. 5.48).

External oppositely directed torques T_1 and T_2 are applied to the disks.

(a)

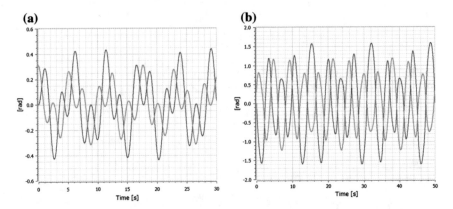

(b)

Fig. 5.45 Examples of dependences of angular displacements on time for two masses (blue is the first mass, yellow is the second mass) for **a** small deviation angles and zero initial velocities, **b** for large deviation angles and nonzero initial velocities

(a)

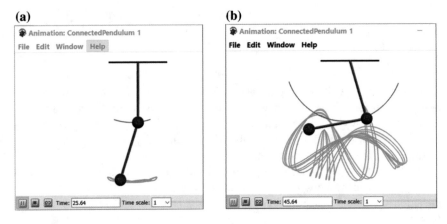

(b)

Fig. 5.46 Examples of animation and tracing of the path of two masses (blue is the first mass, yellow is the second mass) for **a** small deflection angles and zero initial velocities, **b** for large deviation angles and nonzero initial velocities

Physical parameters of the system characterize this system: inertia moments of disks I_1 and I_2, spring stiffness k (torsion modulus), damping constant b (taking into account viscous friction of the medium).

Tasks

Build a model that will describe the change in the angle of displacement of two disks depending on the magnitude of the torques applied to the first and second disks.

Conduct numerical experiments.

Moments of inertia of the disks $I_1 = 0.5$ kg m^2 and $I_2 = 0.5$ kg m^2, spring stiffness $k = 0.1$ kg/s^2 (torsion modulus), attenuation constant $b = 0.01$ N s/m.

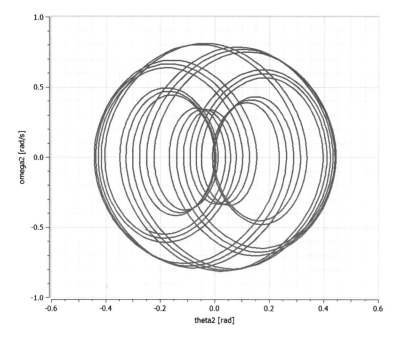

Fig. 5.47 An example of a phase diagram for a single mass of a double pendulum

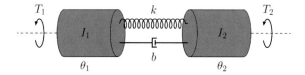

Fig. 5.48 Dual torsion spring oscillator

Apply oppositely directed torques $T_1 = 0.25$ N m and $T_2 = -0.25$ N m to the disks. Carry out a series of experiments for different viscosity of the medium.

With the same system characteristics, apply oppositely directed torques $T_1 = 0.5$ N m and $T_2 = -0.25$ N m that are not equal in modulus. Conduct a series of experiments for different viscosity of the medium.

Modeling and computational experiment

Let θ_1 be the angular displacement of the first disk and θ_2 be the angular displacement of the second disk. We take into account that the returning moments N_1 and N_2 are now proportional to the relative deviation angle:

$$N_1 = -k(\theta_1 - \theta_2),$$
$$N_2 = -k(\theta_2 - \theta_1),$$

The braking torque of viscous friction forces is proportional to the relative angular velocity of each disk $(\dot{\theta}_1 - \dot{\theta}_2)$ or $(\dot{\theta}_2 - \dot{\theta}_1)$.

Given this, we obtain the system of equations:

$$T_1 - b\left(\frac{d\,\theta_1}{dt} - \frac{d\,\theta_2}{dt}\right) - k(\theta_1 - \theta_2) = I_1\frac{d^2\theta_1}{dt^2}$$

$$T_2 - b\left(\frac{d\,\theta_2}{dt} - \frac{d\,\theta_1}{dt}\right) - k(\theta_2 - \theta_1) = I_2\frac{d^2\theta_2}{dt^2}$$

or

$$I_1\ddot{\theta}_1 + b(\dot{\theta}_1 - \dot{\theta}_2) + k(\theta_1 - \theta_2) = T_1$$

$$I_2\ddot{\theta}_2 + b(\dot{\theta}_2 - \dot{\theta}_1) + k(\theta_2 - \theta_1) = T_2$$

The program code is shown in Fig. 5.49.

Apply equal oppositely directed torques $T_1 = 0.25$ and $T_2 = -0.25$ to the disks. In this case, we will observe antiphase oscillations of the disks, damping with increasing viscosity of the medium. During critical attenuation, the disks will rotate in opposite directions without attenuation. The angles of rotation of the disks can be determined numerically from the graphs (Fig. 5.50).

The phase diagram of a double oscillator is two displaced series of turns of spirals; each series has its own "focus" (Fig. 5.51).

Now let the applied torques be not equal modulo $T_1 = 0.5$ and $T_2 = -0.25$. In this case, the system begins to rotate toward a larger applied torque with little or no fluctuation (Fig. 5.52).

This system is easy to assemble from ready-made components in Wolfram SystemModeler. Let us start with the usual spring torsion pendulum. We will go into the

```
Task1

model Task1
    parameter Real T1(unit = "N m") = 0.25 "torque1";
    parameter Real T2(unit = "N m") = -0.25 "torque2";
    parameter Real I1(unit = "kg m2") = 0.5 "Moment of inertia1";
    parameter Real I2(unit = "kg m2") = 0.5 "Moment of inertia2";
    parameter Real b(unit = "N s/m") = 0.01 "coefficient of viscosity";
    parameter Real k(unit = "N/m") = 0.1 "coefficient of spring stiffness";
    Real omega1(unit = "rad/s", start = 0) "angular velocity1";
    Real teta1(unit = "rad", start = 0) "angular displacement1";
    Real omega2(unit = "rad/s", start = 0) "angular velocity2";
    Real teta2(unit = "rad", start = 0) "angular displacement2";
equation
    der(teta1) = omega1;
    der(teta2) = omega2;
    der(omega1) = (T1 - b * (omega1 - omega2) - k * (teta1 - teta2)) / I1;
    der(omega2) = (T2 - b * (omega2 - omega1) - k * (teta2 - teta1)) / I2;
end Task1;
```

Fig. 5.49 Dual torsion spring oscillator simulation program code

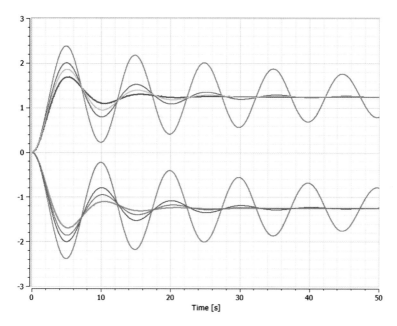

Fig. 5.50 Graphs of the angular displacement of disks with equal angular momentum modulo with different viscosity of the medium

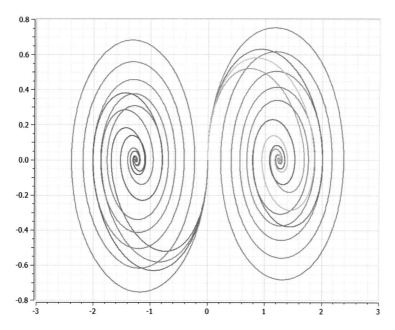

Fig. 5.51 Phase portrait of a damped double torsion oscillator with two "tricks"

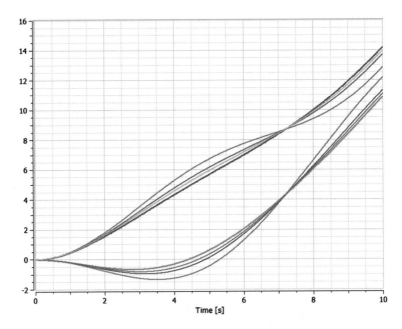

Fig. 5.52 Graphs of the angular displacement of the disks at different angular momentum modulo at different viscosity of the medium

Modelica library and select the Mechanics section there. Now we will be interested in rotational components. Here we select the components of interest to us Spring-Damper (spring with a damper), and Inertia is a component with a given moment of inertia (Fig. 5.53).

In addition to the components, we will need to apply the external torque T to the Inertia component, which we will find in sources. Choose ConstantTorque, i.e., constant torque (Fig. 5.54).

Now everything is ready to create a model. Assemble it in a graphical representation of the selected components (Fig. 5.55).

Carry out similar experiments and make sure that the two approaches give the same results. The model of a double torsion oscillator is also easy to build from ready-made components. The simulation results coincide with the obtained direct numerical solution (Fig. 5.56).

5.7   Gear Rotary Mechanical System

Formulation of the problem

A gear system consists of two torsion spring oscillators that are offset parallel to some *X*-axis. The torsion oscillator disks are coupled to weightless coil springs. Other ends

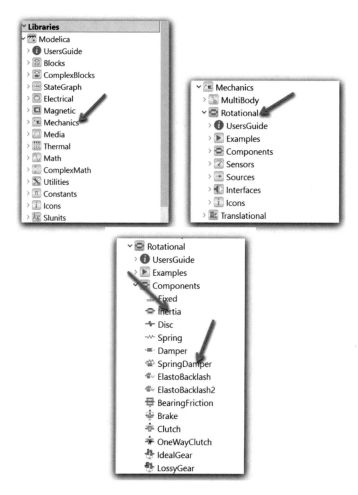

Fig. 5.53 Selection of the necessary components for modeling the torsion oscillator

of the springs are rigidly fixed to gears of different radii r_1 and r_2, coupled to each other (Fig. 5.57).

External torque T is applied to the left disk.

Let I_1, I_2—moments of inertia of the disks, b_1, b_2—constant damping of the left and right springs, respectively, k_1, k_2—spring stiffness, n—gear ratio. As variables, we consider θ_1, θ_2—the angular displacements of the disks

Tasks

Build a model that will describe the change in the displacement angles of two disks in a gear system, depending on the amount of torque applied to the first disk.

Carry out computational experiments with the following data.

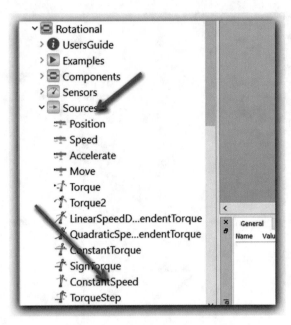

Fig. 5.54 Selection of external torque applied to the system

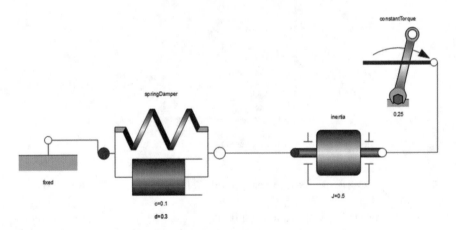

Fig. 5.55 Model of torsion oscillator made from prefabricated components

The disks are the same, with moments of inertia $I = 0.5$ kg m^2, the stiffness of the springs is the same and equal to $k = 0.1$ kg/s^2 (torsion modulus), the damping coefficients of the springs are the same and equal to $b = 0.01$ N s/m. Apply a torque $T = 0.25$ N m to the first disk. Conduct a series of experiments for various gear ratios $n = 0.1; 0.3; 1; 3; 5$.

Apply various torques T to the first disk and investigate the behavior of the system.

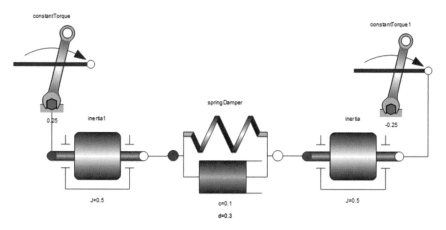

Fig. 5.56 A model of a rotational mechanical system with two moments of inertia made from prefabricated components

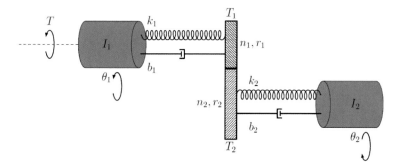

Fig. 5.57 Gear system

Modeling and computational experiment

The gear ratio is the ratio of the number of teeth of the drive gear to the number of teeth of the driven gear. The wheel to which the torque is supplied from outside is called the driving one, and the wheel from which the moment is removed is called the driven one. In this problem, the wheel to which the first disk is attached is the leading one, and the wheel that is connected to the second disk is the driven one. The gear ratio of the gear is determined by:

$$\frac{n_1}{n_2} = n$$

where n_2 is the number of teeth of the driven gear and n_1 is the number of teeth of the drive wheel. The number of teeth is proportional to the radius of the wheel, whence we have:

$$\frac{r_1}{r_2} = \frac{n_1}{n_2} = n$$

The angular displacement is related to the radii as follows:

$$r_1\theta_1 = r_2\theta_2$$

from where $\theta_2 = n\theta_1$.

We examined the geometrical characteristics of the gear train. However, the gear ratio is the main kinematic and dynamic characteristic of the gear transmission. This is the ratio of the angular velocity of the first wheel to the angular velocity of the second wheel.

$$\frac{d\,\theta_2}{dt} \Big/ \frac{d\,\theta_1}{dt} = n$$

The gear ratio shows how many times the angular speed has decreased, and the torque has increased by the same amount (without taking into account losses). Indeed, equating the power transmitted by the gear transmission, we obtain:

$$T_1\frac{d\,\theta_1}{dt} = T_2\frac{d\,\theta_2}{dt}$$

from where

$$\frac{T_1}{T_2} = n$$

If the diameter of the drive wheel is smaller, then the torque of the driven wheel increases due to a proportional decrease in rotation speed. In accordance with the gear ratio (this is the ratio of the rotational speed of the leading element of a mechanical transmission to the rotational speed of the driven one), an increase in torque will cause a proportional decrease in the angular speed of rotation of the driven wheel, and their product—mechanical power—will remain unchanged. Simply put, if small gears drive large gears, the torque increases, and vice versa, if large gears drive small gears, the torque decreases.

To compile a mathematical model of the system, we consider the torques of the wheels

$$T_1 = T - I_1\frac{d^2\theta_1}{dt^2} - b_1\frac{d\,\theta_1}{dt} - k_1\theta_1$$

$$T_2 = I_2\frac{d^2\theta_2}{dt^2} + b_2\frac{d\,\theta_2}{dt} + k_2\theta_2$$

```
Task2                                                                    model    Modelica Text View
model Task2
  parameter Real T(unit = "N m") = 0.25 "torque";
  parameter Real I1(unit = "kg m2") = 0.5 "Moment of inertia1";
  parameter Real I2(unit = "kg m2") = 0.5 "Moment of inertia2";
  parameter Real b1(unit = "N s/m") = 0.01 "coefficient of viscosity1";
  parameter Real b2(unit = "N s/m") = 0.01 "coefficient of viscosity2";
  parameter Real k1(unit = "N/m") = 0.1 "coefficient of spring stiffness1";
  parameter Real k2(unit = "N/m") = 0.1 "coefficient of spring stiffness2";
  parameter Real n = 3;
  Real omega1(unit = "rad/s", start = 0) "angular velocity1";
  Real teta1(unit = "rad", start = 0) "angular displacement1";
  Real omega2(unit = "rad/s", start = 0) "angular velocity2";
  Real teta2(unit = "rad", start = 0) "angular displacement2";
  equation
  der(teta1) = omega1;
  der(teta2) = omega2;
  teta2 = n * teta1;
  der(omega1) = (T - (b1 + b2 * n ^ 2) * der(teta1) - teta1 * (k1 + k2 * n ^ 2)) / (I1 + I2 * n ^ 2);
  ;
end Task2;
```

Fig. 5.58 Gear simulation program code

Given that

$$T_1 = nT_2$$

$$\theta_2 = n\theta_1$$

We see that this system is reduced to one equation:

$$\left(I_1 + n^2 I_2\right)\frac{d^2\theta_1}{dt^2} + \left(b_1 + n^2 b_2\right)\frac{d\theta_1}{dt} + \left(k_1 + n^2 k_2\right)\theta_1 = T$$

The program code is shown in Fig. 5.58.

We will carry out two series of experiments. In the first, we will consider the external moment constant and equal to $T = 0.25$ N m. When the value of n changes, the ratio of the speeds of the first and second disks changes. For $n > 1$ (the first wheel is larger than the second), the rotation speed increases (upper graph, green line), for $n < 1$ (the first wheel is smaller than the second), the rotation speed decreases (second graph, green line), and finally, with equal wheel sizes $n = 1$, the speeds are equal ($n = 0.99$ is taken on the lower graph for clarity) (Fig. 5.59).

Let us move on to the second series of experiments. We fix the gear with the number $n = 3$ and apply an external torque $T = 0.25$ N m (yellow and brown graphs), apply an oppositely directed torque $T = -0.25$ N m—the rotation speed changes in antiphase compared to the first case (red and green graphs)—and, finally, increase the torque $T = 0.5$ N m—the speed has increased, as can be seen from the graphs (blue and violet) (Fig. 5.60)

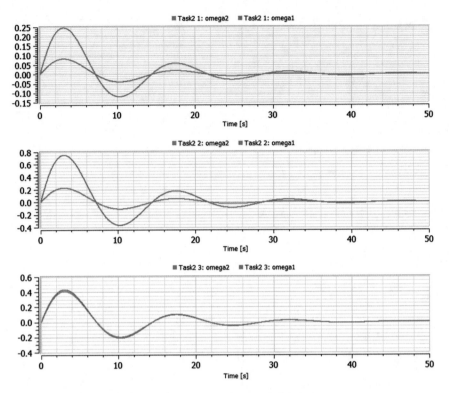

Fig. 5.59 Angular velocity of the disks at gear ratios $n = 3; 0.3; 1$

5.8 Mechanical System with Two Springs and a Block

Formulation of the problem

Let the mechanical system consists of two ideal springs with individual stiffness k_1 and k_2, fixed on independent suspensions, and an inextensible thread is attached to the ends of the springs, on which the mass block m_p is located. A block of mass m_L can be suspended from the block.

The positive directions of the linear velocities of the springs v_1, v_2, the vertical speed of the block v_3, and its angular velocity ω are shown in Fig. 5.61.

Tasks

It is necessary to study this system, making up equations describing the change in the system over time, and build its computer model under various initial conditions. Investigate the behavior of the system depending on the initial conditions themselves and the way they are set.

Initial conditions: spring extension and block speeds v_3 and w are equal to zero, $k_1 = 5$ N/m, $k_2 = 10$ N/m, $m_p = 0.5$ kg, $m_L = 10$ kg, $R_p = 0.3$ m. Initial values of state variables are set using start attribute.

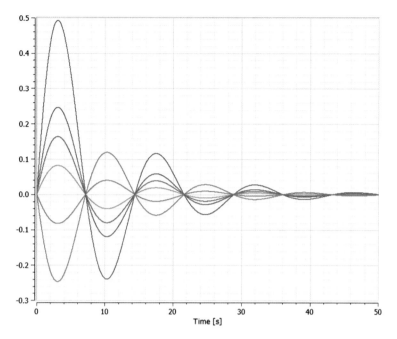

Fig. 5.60 Angular speeds of disks at different applied moments

Fig. 5.61 Mechanical
system with two springs and
a block

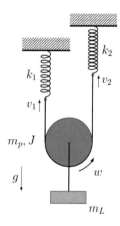

Modify the original model, assuming that the system is initially in a stable state. The values of the system parameters remain the same as in the previous task. The mass of the load $m_L = 10$ kg, 5 s after the start of the experiment, the load is removed from the disk. To fulfill the initial conditions when writing code, use the initial equations initial equation to specify the initial values of the derivatives.

Modify the original model, assuming that the model is in a stable state when the initial spring extensions are 3 and 1.5 m. Find new system parameters.

Modeling and computational experiment

Let the block be a continuous round disk of radius R_p, then its moment of inertia I can be calculated by the formula:

$$I = \frac{1}{2} m_p R_p^2$$

The behavior of ideal springs obeys Hooke's law $F_i = k_i e_i$, where e_i is their elongation (the difference between the actual and undeformed length of the spring). Knowing the elongation of the springs, you can find their linear speeds:

$$\frac{de_i}{dt} = -v_i$$

Let the load be suspended on the block; then according to Newton's second law for linear acceleration of the block:

$$(m_p + m_L) \frac{dv_3}{dt} = k_1 e_1 + k_2 e_2 - g(m_p + m_L)$$

When the unit rotates, the springs attached to it create a returning moment of force N, proportional to the shoulder R_p and the resultant applied forces:

$$N = R_p(k_2 e_2 - k_1 e_1),$$

Applying the basic equation of the dynamics of rotation of a rigid body about a fixed axis to the motion of a disk with a moment of inertia I,

$$N = I \frac{d\omega}{dt} = I \frac{d^2\theta}{dt^2}$$

We get

$$I \frac{d\omega}{dt} = R_p(k_2 e_2 - k_1 e_1)$$

The displacement of the springs x_1, x_2, the vertical displacement of the block x_3, and the angle θ of rotation of the block are connected as follows:

$$x_1 = x_3 - R_p\theta$$
$$x_2 = x_3 + R_p\theta$$

Differentiating these two equations in time, we obtain:

$$v_1 = v_3 - R_p\omega$$
$$v_2 = v_3 + R_p\omega$$

Let us move on to writing code to build the model. In the first experiment, we set the initial conditions as follows:

```
    SI.Length e1(start = 0, fixed = true);
// initial lengthening of the first spring
    SI.Length e2(start = 0, fixed = true);
// initial lengthening of the second spring
    SI.Velocity v3(start = 0, fixed = true);
// initial  linear disk speed
    SI.AngularVelocity w(start = 0, fixed = true);
// initial angular velocity of the disk
```

We set them rigidly, i.e., the system is in the neutral position at the initial moment of time, after which it is released without a push.

The complete code for this model is shown in Fig. 5.62.

Let us perform the first experiment. In it, at the initial moment of time, the springs are in an undeformed state; therefore, the system will begin to shift downward under the action of gravity, as a result of which we observe an oscillatory process. Let us start with the study of elongations (Fig. 5.63).

At the first moment of time, the elongations are equal to zero, after which they begin to increase, and the spring stretches with greater rigidity less. Oscillations occur in a vacuum and are harmonious. Observe the change in speed. The change in the vertical speed of the disk is harmonic (blue line), while beats (yellow and green lines) are observed for the speed of extension of the springs, as shown in Fig. 5.64.

The change in the angular velocity of the disk is also inharmonic and is shown in Fig. 5.65.

Let us perform the second experiment. In it, the system at the initial moment of time should be in a stable state, i.e., at rest. The speed of the springs should be zero. Therefore, we use the initial equation to set the initial values of the speeds of the springs and the disk:

```
initial equation
der(e1) = 0; // initial speed of the first spring
der(e2) = 0; // initial speed of the second spring
der(v3) = 0; // initial linear disk speed
der(w) = 0; // initial angular velocity of the disk
```

It is easy to understand that the solution of such a system of equations will give us the lower position of the system as the initial one. However, displacement from the lower position with a given load mass of 10 kg is impossible. The system is stably at rest. This is easy to verify both analytically and experimentally. Therefore, the task

```
SystemWithPulley1
    //connect the library of constants
    constant SI.Acceleration g = 9.81;
    // Spring parameters:
    parameter SI.TranslationalSpringConstant k1 = 5;
    parameter SI.TranslationalSpringConstant k2 = 10;
    // Pulley parameters
    parameter SI.Mass Mp = 0.5;
    parameter SI.Radius Rp = 0.3;
    parameter SI.MomentOfInertia J = 0.5 * Mp * Rp ^ 2;
    // Mass parameters
    parameter SI.Mass ML = 10;
    //time dependent parametres:
    SI.Length e1(start = 0, fixed = true);
    SI.Length e2(start = 0, fixed = true);
    SI.Velocity v1;
    SI.Velocity v2;
    SI.Velocity v3(start = 0, fixed = true);
    SI.Force F1;
    SI.Force F2;
    SI.AngularVelocity w(start = 0, fixed = true);
equation
    // for spring 1:
    F1 = k1 * e1;
    der(e1) = -v1;
    // for spring 2:
    F2 = k2 * e2;
    der(e2) = -v2;
    // for mass and pulley:
    (Mp + ML) * der(v3) = F1 + F2 - g * (Mp + ML);
    v1 = v3 - Rp * w;
    v2 = v3 + Rp * w;
    J * der(w) = Rp * (F2 - F1);
    //Modelling time
    □;
end SystemWithPulley1;
```

Fig. 5.62 Program code of a mechanical system with two springs and a block with zero initial conditions (attributes of state variables)

requires you to remove the load after 5 s of the experiment, which will bring our system out of balance.

Here it is necessary to pay attention to the fact that an additional condition for instant removal of cargo turns our model into a hybrid one. Take this into account when writing code.

```
parameter SI.Mass ML0 = 10;
SI.Mass ML;
ML = if time < 5 then ML0 else 0;
(Mp + ML)*der(v3) = F1 + F2 - g*(Mp + ML);
```

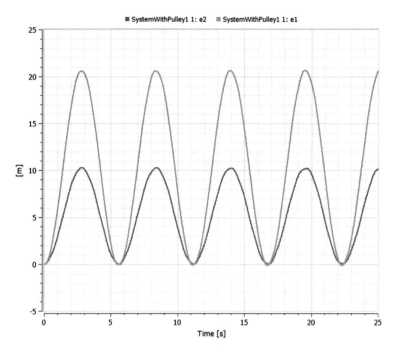

Fig. 5.63 Schedule of change in time of elongation of two springs (e_1—for the first spring, e_2—for the second)

The full code for this model is given below.

Addition: full model code of a mechanical system with two springs and a block with zero initial equations (zero values of the derivatives of the displacements of the springs and the block)

```
model SpringPulleyLoad2
import SI = Modelica.SIunits;
constant SI.Acceleration g=9.81; //Connect the library
of Constants Models
// Spring Parameters:
parameter SI.TranslationalSpringConstant k1=5;
parameter SI.TranslationalSpringConstant k2=10;
// Block Parameters
parameter SI.Mass Mp=0.5;
parameter SI.Radius Rp=0.3;
parameter SI.MomentOfInertia J=0.5*Mp*Rp^2;
// Load Parameters
parameter SI.Mass ML0=10;
//Set Time-Dependent Variables:
SI.Mass ML;
SI.Length e1;
```

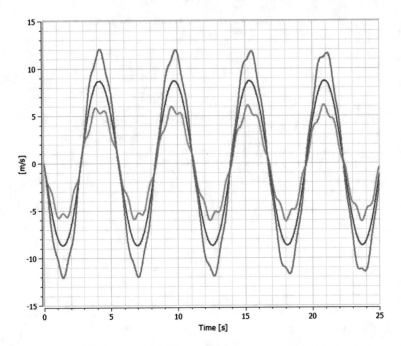

Fig. 5.64 Graph of the time variation of the tensile speeds of the springs (v_1—for the first spring, v_2—for the second) and the vertical speed of the disk v_3

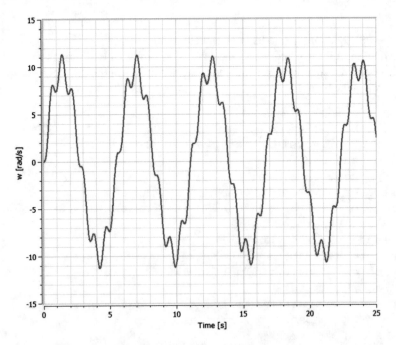

Fig. 5.65 Graph of the time variation of the disk angular velocity w

```
SI.Length e2;
SI.Velocity v1;
SI.Velocity v2;
SI.Velocity v3;
SI.Force F1;
SI.Force F2;
SI.AngularVelocity w;
equation
// For spring 1:
F1 = k1*e1;
der(e1) = -v1;
// For spring 2:
F2 = k2*e2;
der(e2) = -v2;
// For block and load:
ML = if time < 5 then ML0 else 0;
(Mp + ML)*der(v3) = F1 + F2 - g*(Mp + ML);
v1 = v3 - Rp*w;
v2 = v3 + Rp*w;
J*der(w) = Rp*(F2 - F1);
// Initial equations
initial equation
der(e1) = 0;
der(e2) = 0;
der(v3) = 0;
der(w) = 0;
annotation(experiment(StartTime=0,StopTime=25));
//Specify the simulation time
end SpringPulleyLoad2;
```

The behavior of the hybrid model is reflected in the graphs Figs. 5.66, 5.67, and 5.68 where in the first 5 s the system is in lower stable equilibrium, after which an event occurs—the load is removed from the unit and the system begins to oscillate.

When performing the third task, a problem arises related to the fact that the system must be in equilibrium at certain spring extensions. This can be achieved by changing the stiffness of the springs. You can calculate the required stiffness analytically, but you can do differently and provide the solver with the opportunity to select them in the process of a computational experiment. To do this, we calculate the position of stable equilibrium using the initial equations as well as was done in the previous task, with fixed values of elongations:

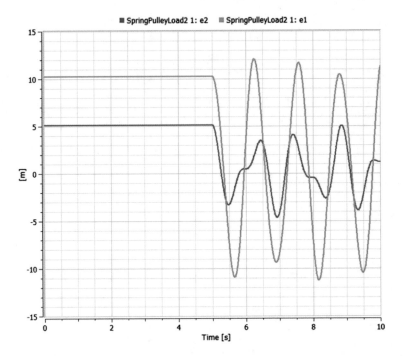

Fig. 5.66 Schedule of change in time of elongation of two springs (e_1—for the first spring, e_2—for the second)

```
initial equation
der(e1) = 0; // initial speed of the first spring
der(e2) = 0; // initial speed of the second spring
der(v3) = 0; // initial linear disk velocity
der(w) = 0; // initial angular velocity of the disk
e1 = 3; // initial elongation of the first spring
e2 = 1.5; // initial elongation of the second spring
```

The value of the start attribute is the value as the intended value. Valid values for these parameters must be specified.

```
     parameter SI.TranslationalSpringConstant
k1(start=1,fixed=false);
     parameter SI.TranslationalSpringConstant
k2(start=1,fixed=false);
```

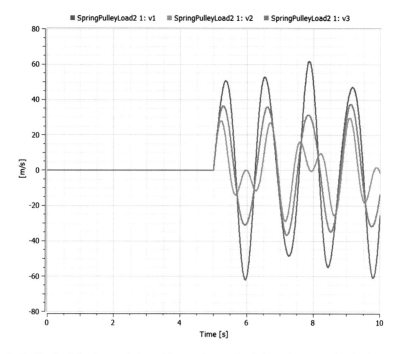

Fig. 5.67 Graph of the time variation of the tensile speeds of the springs (v_1—for the first spring, v_2—for the second) and the vertical speed of the disk v_3

The full code for this model is given below.

Addition: full model code of a mechanical system with two springs and a block with zero initial equations (zero values of the derivatives of the displacements of the springs and the block) and the specified displacements of the springs

```
model SpringPulleyLoad3
import SI = Modelica.SIunits; //Connect the library of
Constants Models
constant SI.Acceleration g=9.81;
// Spring Parameters:
parameter SI.TranslationalSpringConstant
k1(start=1,fixed=false);
parameter SI.TranslationalSpringConstant
k2(start=1,fixed=false);
// Block Parameters
parameter SI.Mass Mp=0.5;
parameter SI.Radius Rp=0.3;
parameter SI.MomentOfInertia J=0.5*Mp*Rp^2;
```

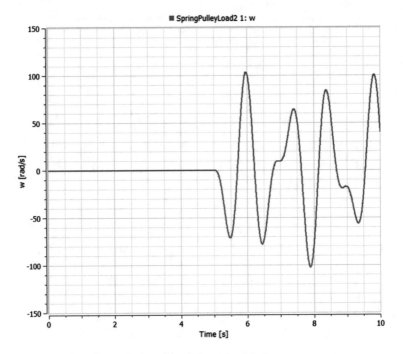

Fig. 5.68 Graph of the time variation of the disk angular velocity *w*

```
// Load Parameters
parameter SI.Mass ML0=10;
// Set Time-Dependent Variables:
SI.Mass ML;
SI.Length e1;
SI.Length e2;
SI.Velocity v1;
SI.Velocity v2;
SI.Velocity v3;
SI.Force F1;
SI.Force F2;
SI.AngularVelocity w;
equation
// For the spring 1:
F1 = k1*e1;
der(e1) = -v1;
// For the spring 2:
F2 = k2*e2;
der(e2) = -v2;
```

```
// For block and load:
ML = if time < 5 then ML0 else 0;
(Mp + ML)*der(v3) = F1 + F2 - g*(Mp + ML);
v1 = v3 - Rp*w;
v2 = v3 + Rp*w;
J*der(w) = Rp*(F2 - F1);
// Initial equations
initial equation
der(e1) = 0;
der(e2) = 0;
der(v3) = 0;
der(w) = 0;
e1 = 3;
e2 = 1.5;
annotation(experiment(StartTime=0,StopTime=25));
//Specify the simulation time
end SpringPulleyLoad3;
```

Figure 5.69 shows a comparison between the second and third experiments. It can be seen from this comparison that the solver increased the stiffness of the springs in

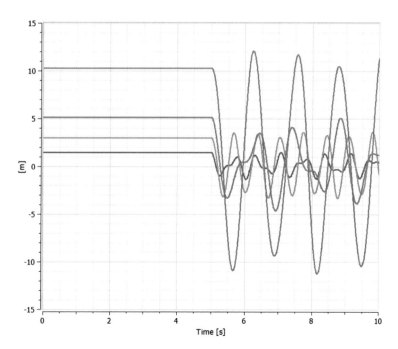

Fig. 5.69 Graphs of changes in time of elongation of two springs (e_1—for the first spring, e_2—for the second). Red and green graphs—the second experiment, blue and yellow graphs—the third experiment

order to achieve the fulfillment of the set initial conditions. As a result, the amplitude of the oscillations decreased (blue and yellow graphs).

The stiffness found as a result of the computational experiment turned out to be $k_1 = 17.1675$ N/m and $k_2 = 34.335$ N/m.

5.9   Sophisticated Mechanical System with Springs and Block

Formulation of the problem

We consider a mechanical system consisting of two springs with individual stiffness k_1 and k_2, a damper with a coefficient b_1, a weightless pulley and lever (rotating board), fulcrum. The first spring begins to move and moves according to the periodic law for 20 s. Speed v_1 is a known function of time:

$$v_1(t) = \begin{cases} V_{10} \cdot \sin(\omega t), & \text{if } 2 < t < 20 \\ 0 \end{cases}$$

where v_{10} and ω are known parameters.

The distances from the fulcrum to the points of application of forces to the board are known and equal to L_2, L_3, and L_4, respectively. The velocities of these points are denoted by v_2, v_3, and v_4. The directions of the arrows in Fig. 5.70 indicate the direction of shift of the lever.

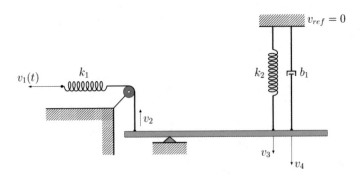

Fig. 5.70 Sophisticated mechanical system with springs and block

Tasks

It is necessary to study this system, making equations describing the change in the system over time, and build its computer model.

Write equations describing the change in elongation time and spring speeds.

Conduct a simulation and computational experiment in a WSM environment.

Modeling and computational experiment

When modeling the system, we make the following simplifying assumption. We assume that the force applied to the left end of the lever is instantly balanced by the forces applied to the right of the fulcrum, i.e., the elasticity of the second spring and the force of viscous friction. Under this assumption, the lever is in equilibrium at each instant of time, i.e., the rule of moments is fulfilled for him:

$$k_1 e_1 \cdot L_2 = k_2 e_2 \cdot L_3 + b_1 v_4 \cdot L_4$$

Here e_1 and e_2 are the extensions of the springs. In addition, at each point of the lever, its angular velocity is the same:

$$\omega = \left| \frac{v_2}{L_2} \right| = \left| \frac{v_3}{L_3} \right| = \left| \frac{v_4}{L_4} \right|$$

The module in this formula indicates that linear velocities v_2, v_3, and v_4 may have different signs.

The spring extension speeds are related to the difference in the speeds of the points of their attachment:

$$\begin{cases} \frac{de_1}{dt} = v_1 - v_2 \\ \frac{de_2}{dt} = v_3 \end{cases}$$

The above relations make up the mathematical model of the problem (Fig. 5.71).

Let us conduct a numerical experiment. We plot the velocities in the system. It can be seen that with external periodic exposure, oscillations are excited in the system, which cease when the exposure ceases (Fig. 5.72).

According to the schedule, it is possible to analyze the fluctuations of the lever—the right end of the lever moves periodically with a constant offset relative to external influences. The left end of the lever is out of phase with the right.

Here are the spring extension graphs (yellow graph—left spring, blue graph—right spring) (Fig. 5.73).

```
┌─────────────────────────────────────────────────────────────────────┐
│ ▣                           SpringsDamperLever                    ▣  │
├─────────────────────────────────────────────────────────────────────┤
│ SpringsDamperLever                                                    │
├─────────────────────────────────────────────────────────────────────┤
│ model SpringsDamperLever                                              │
│   import SI = Modelica.SIunits;                                       │
│   parameter SI.Velocity V10 = 0.01;                                   │
│   parameter SI.AngularFrequency w = 1;                                │
│   parameter SI.Length L2 = 1;                                         │
│   parameter SI.Length L3 = 2;                                         │
│   parameter SI.Length L4 = 3;                                         │
│   parameter SI.TranslationalSpringConstant k1 = 5;                    │
│   parameter SI.TranslationalSpringConstant k2 = 10;                   │
│   parameter SI.TranslationalDampingConstant b1 = 5;                   │
│   SI.Length e1(start = 0, fixed = true);                              │
│   SI.Length e2(start = 0, fixed = true);                              │
│   SI.Velocity v1;                                                     │
│   SI.Velocity v2;                                                     │
│   SI.Velocity v3;                                                     │
│   SI.Velocity v4;                                                     │
│   SI.Force F2;                                                        │
│   SI.Force F3;                                                        │
│   SI.Force F4;                                                        │
│ equation                                                              │
│   v1 = if time > 2 and time < 20 then V10 * sin(w * time) else 0;     │
│   F2 = k1 * e1;                                                       │
│   der(e1) = v1 - v2;                                                  │
│   F2 * L2 + F3 * L3 + F4 * L4 = 0;                                    │
│   v2 / L2 = -v3 / L3;                                                 │
│   v2 / L2 = -v4 / L4;                                                 │
│   F3 = -k2 * e2;                                                      │
│   der(e2) = v3;                                                       │
│   F4 = -b1 * v4;                                                      │
│ end SpringsDamperLever;                                               │
└─────────────────────────────────────────────────────────────────────┘
```

Fig. 5.71 Program code for a model of a complex mechanical system with springs and a block

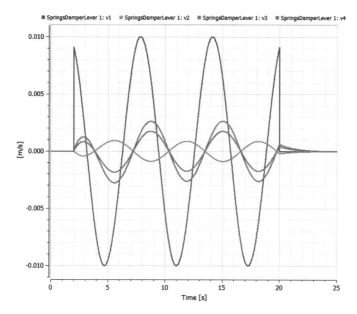

Fig. 5.72 Graph of the time variation of the tensile speeds of the springs—the blue graph is the speed of the left spring as a result of external influences, the red and green graphs are the speeds of the points to the right of the support, and the yellow graph is the speed of the left end of the lever

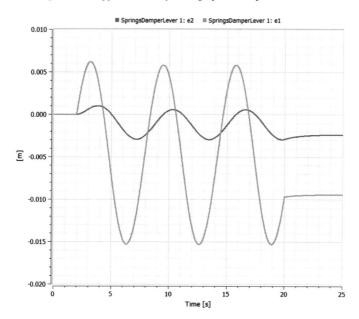

Fig. 5.73 Graphs of changes in time of elongation of two springs (e_1—for the first spring, e_2—for the second)

References

1. V.A. Ryzhov, T.A. Fedorova, K.S. Safronov, N.V. Tryaskin, Virtualnye laboratornye raboty v srede Wolfram SystemModeler. Uchebnoe posobie. – Izdatelstvo SPbGMTU, SPb (2019)
2. V.A. Ryzhov, T.A. Fedorova, K.S. Safronov, N.V. Tryaskin, Rukovodstvo po vypolneniju virtualnyh laboratornyh rabot v srede Wolfram SystemModeler. Uchebnoe posobie. – Izdatelstvo SPbGMTU, SPb (2019)

Chapter 6
Hierarchical Component Models

The basis of any modeling system is a block, an analog of a real device or a component. Blocks can be interconnected, forming functional diagrams located on the plane. Any functional diagram can be considered as a complex unit. These complex blocks with internal structure we can again interconnect building hierarchical multi-level systems [1–3].

A block is an independent element that functions according to its internal laws and interacts with the outside world (other blocks) through a given set of variables called interface variables. A block is usually drawn in the form of a rectangle, depicting interface variables of various types next to it, for example, these variables can be "inputs-outputs" or "contacts" (see Fig. 6.1). The type of interface variables determines the way to automatically build an aggregate system according to a given description of individual blocks and their relationships.

We will distinguish between the task of constructing a classification of the components from which the future system is built and the task of constructing a specific system from the available components. In the world of real physical objects, there is a set of standard components and many devices in which they are used as prefabricated elements. In modeling packages, class libraries correspond to them, and block-functional diagrams of designed devices built from instances of existing classes. In any functional diagram, one can distinguish between the types of blocks (classes) used for its construction and specific blocks (instances of classes).

Fig. 6.1 Block with inputs and outputs and block with contacts

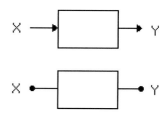

K. Rozhdestvensky et al., *Computer Modeling and Simulation of Dynamic Systems Using Wolfram SystemModeler*, https://doi.org/10.1007/978-981-15-2803-3_6

The presence of a connection between contacts means that the values of the variables corresponding to the contacts are equal at any time. In modern visual packages, there are two kinds of connections:

(1) Unidirectional (oriented), and then, the connected contacts are divided into a receiver and a source, and it is also postulated that the receiver cannot influence the source.

The oriented block is used for the design of input-state-output systems. The use of oriented blocks involves the introduction of restrictions on the use of phase vector variables in the preparation of equations describing the behavior.

(2) Bidirectional (non-oriented), in this case, the connected contacts are equal. The idea to use non-oriented blocks as components arose a very long time ago and is especially vivid in the design of electrical circuits.

The development of this approach is the connection through the contacts of blocks containing a description of behavior in the form of systems of algebraic and differential equations. The main idea of the approach is very well expressed in the definition of "language for modeling physical systems." When using the traditional block modeling approach to describe a number of real systems, serious restrictions arise on the type of blocks.

This chapter is devoted to building hierarchical models and creating your own libraries in the SystemModeler package. First, a model of a tank with a flat bottom is developed, then a similar model based on components is created. We will also be able to appreciate the flexibility of this approach, allowing us to test new scenarios.

6.1 Draining of a Tank

Formulation of Problem 1

This problem is adopted from [4]. There is a tank with a flat bottom, as shown in Fig. 6.2, into which a liquid of density ρ enters at a constant speed q_i and leaves at a speed q_o through an opening in the lower part of the tank. The area of the base of the tank is A. The height of the liquid in the vessel is a function of time $h(t)$.

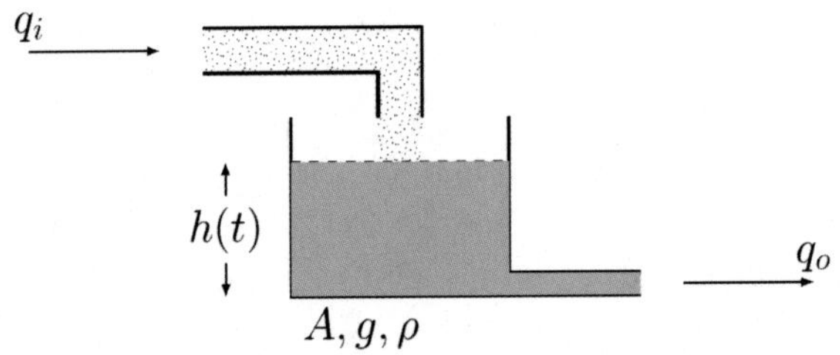

Fig. 6.2 Scheme of fluid intake in the tank

Tasks

Build a mathematical model of fluid flow into the tank based on the principle of equivalence of mathematical models. Derive the differential equation for the liquid level in the tank. Show the relationship between the height h of the liquid and the input flow rate q_i.

Formulation of Problem 2

Consider the reservoir from the previous example with the base area $A = 1\,\text{m}^2$. Let the input stream be constant for the first 150 s, after which it increases three times:

$$q_i = \begin{cases} f, & t < 150\,\text{s} \\ 3f, & t \geq 150\,\text{s} \end{cases}$$

where f is the given parameter. When performing the simulation, we set $f = 0.15\,\text{m}^3/\text{s}$. And finally, we will limit the flow at the outlet of the tank to the minimum $V_{min} = 0\,\text{m}^3/\text{s}$ and maximum $V_{min} = 0.5\,\text{m}^3/\text{s}$.

Tasks

Controlling the output stream with the help of a proportional-integral controller (PI controller) with an integrated gain $K = 0.1\,\text{m}^2/\text{s}$ and a time constant $T = 10\,\text{s}$, it is required to maintain the tank filling level at a height $h_{ref} = 0.25\,\text{m}$.

Replace the PI controller with the PID controller and compare the resulting models. You need the value of the time constant of the differential gain, take it equal to $T_d = 5\,\text{s}$.

Formulation of Problem 3

Consider a system consisting of three identical reservoirs, with a fluid flow controlled by three PI regulators. Suppose that the level $h_{ref} = 0.2\,\text{m}$ is required to be maintained in the first and third tanks, and the level increase to $h_{ref} = 0.4\,\text{m}$ is permissible in the second tank.

Tasks

Build a hierarchical model of the system using the components created in the previous tasks.

Modeling and computational Experiment 1

We construct a mathematical model of fluid intake into the tank based on the principle of equivalence of hydraulic systems and electrical circuits. The main elements of hydraulic systems are hydraulic resistance, hydraulic capacity, and inertia (see Fig. 6.3).

These elements are similar to their electrical equivalents—resistance, capacitance, and inductance. The electric current in this approach is equivalent to the fluid flow

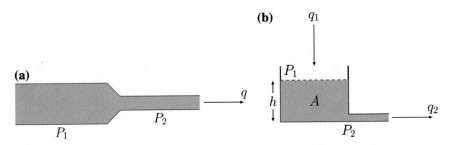

Fig. 6.3 Main elements of the hydraulic system

rate in the system, and the potential difference in the electrical circuits is similar to the pressure difference in hydraulic systems.

Hydraulic capacity is a measure of energy conservation in a hydraulic system. An example of a hydraulic reservoir is a reservoir in which potential energy is stored. Consider the tank shown in Fig. 6.3b. If q_1 and q_2 are the inflowing and outflowing flows, respectively, and V is the volume of fluid inside the reservoir, then we can write the differential equation

$$q_1 - q_2 = \frac{dV}{dt} = A\frac{dh}{dt}$$

Here, A is the area of the base of the tank, and h is the height of the liquid column in the tank. The pressure difference at the upper and lower levels of the tank is determined by the expression:

$$p = P_1 - P_2 = h\rho g$$

where the expression for the height of the liquid level is given by:

$$h = \frac{p}{\rho g}$$

Substituting the expression for the height in the differential equation, we finally obtain:

$$q_1 - q_2 = \frac{A}{\rho g}\frac{dp}{dt} = C\frac{dp}{dt}$$

where $C = \frac{A}{\rho g}$ is the hydraulic capacity.

Modeling and computational Experiment 2

Let us go back to the original differential equation and rewrite it as:

$$\frac{dh}{dt} = \frac{q_1 - q_2}{A}$$

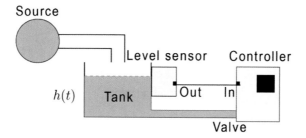

Fig. 6.4 Graphical representation of a tank system

Add to the circuit shown in Fig. 6.2 management system. We will control the liquid level h in the tank using a PI controller, i.e., a controller with proportional and integral regulation, as shown in Fig. 6.4.

The controller calculates the error signal, defined as the difference between the set value and the actual value of the controlled variable. The output value of the controller is calculated as the sum of two variables: one directly proportional to the error signal and the other proportional to the integral of the error signal.

Let $e(t)$ be the control of the error signal, i.e., the difference between the set and the actual liquid level determined by the sensor, h_{ref} is the set liquid level, $h(t)$ is the current liquid level in the tank, and $u_h(t)$ is the output signal that controls the valve position.

Then, the PI controller is described by the following equations:

$$e(t) = h_{\text{ref}} - h(t)$$

$$u_h(t) = K_p \cdot e(t) + \frac{1}{T_I} \cdot \int_0^t e(t) \cdot dt$$

where K_p is the gain of the proportional component (proportional to the PI controller parameter), and T_I is the integration constant (integral parameter of the PI controller), a value that represents the time interval during which the integral component of the output value reaches the input value. To create code in Modelica, we have to get rid of the integral component. To do this, we introduce the notation for the error signal integral

$$I(t) = \int_0^t e(t) \cdot dt$$

Then, the equations will take the following differential form:

$$e(t) = h_{\text{ref}} - h(t)$$

$$\frac{dI(t)}{dt} = e(t)$$

$$u_h(t) = K_p \cdot \left(e(t) + \frac{1}{K_p T_I} \cdot I(t) \right)$$

We introduce the following convenient notation:

$$\frac{1}{K_p T_I} \cdot I(t) = x(t), \quad K_p T_I = T, \quad K_p = K$$

Finally, the equations are rewritten as:

$$\frac{dx(t)}{dt} = \frac{1}{T} e(t)$$
$$u_h(t) = K \cdot (e(t) + x(t))$$

Here, $x(t)$ is the state variable of the PI controller, T is the time constant of the PI controller, and K is the gain of the PI controller. The last two equations will be used when writing code.

We will write the code in *Modelica* in the same way as we did in the previous sections. The only difference is that we will create and use the *LimitValue* function, which will limit the output stream. Its definition is done by creating a function in a package called *Functions*. Since this is a package, when creating a class, we select the class type *Package* (Fig. 6.5).

When creating a class inside a package, you must select the type of the *Function* class (Fig. 6.6).

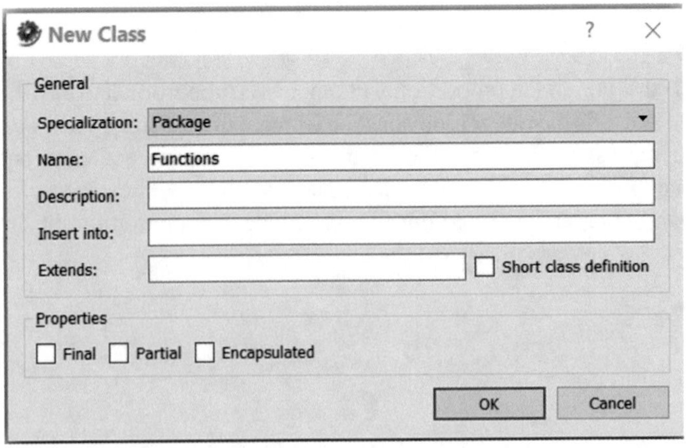

Fig. 6.5 Creating a special *Functions* package for helper functions

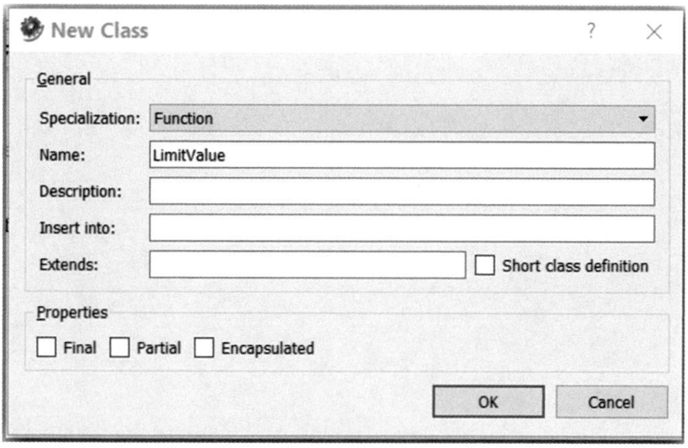

Fig. 6.6 Creating a *LimitValue Function* (output limit)

This function, which has three input variables and one output variable, is as follows:

```
function LimitValue
    input Real pMin;
    input Real pMax;
    input Real p;
    output Real pLim;
algorithm
    pLim:=if p > pMax then pMax else if p < pMin then pMin
else p;
end LimitValue;
```

Now, we will create the main model using the obtained equations and parameter values given to us in the task:

```
model FlatTank "Flat model of a tank"
    parameter Real q0(unit = "m3/s") = 0.15 "Scaling of input
flow";
    parameter Real A(unit = "m2") = 1.0 "Tank bottom area";
    parameter Real K(unit = "m2/s") = 0.1 "PI controller gain";
    parameter Real T(unit = "s") = 10 "PI controller integrator
time constant";
    parameter Real minFlow(unit = "m3/s") = 0.0 "Minimum flow
through output valve";
    parameter Real maxFlow(unit = "m3/s") = 0.5 "Maximum flow
through output valve";
    parameter Real ref(unit = "m") = 0.25 "Reference level
for control";
    Real h(start = 0, unit = "m") "Tank level";
    Real qIn(unit = "m3/s") "Flow through input valve";
    Real qOut(unit = "m3/s") "Flow through output valve";
    Real qOutMax(unit = "m3/s") "Maximum output flow
considering that the tank level cannot be negative";
```

```
    Real u(unit = "m3/s") "Output flow demanded by controller";
    Real e(unit = "m") "Deviation from reference level";
    Real x(unit = "m") "State variable for controller";
equation
    assert(minFlow >= 0, "minFlow - minimum flow through
output valve must be >= 0");
    der(h) = (qIn - qOut) / A;
    qIn = if time > 150 then 3 * q0 else q0;
    qOutMax = if h > 0 then maxFlow else min(qIn, maxFlow);
    qOut = if (-u) < minFlow then minFlow elseif (-u) >
qOutMax then qOutMax else -u;
    e = ref - h;
    der(x) = e / T;
    u = K * (e + x);
end FlatTank;
```

Before performing a numerical experiment, we switch to the icon creation mode and create a conditional image of the tank. To do this, we will use the *Ellipse* form. The upper ellipse is flat, and the three lower ones are depicted according to the pattern of a vertical cylinder, which will give them the correct volume. The selection of appropriate shape (*Ellipse*) and sample (*Vertical Cylinder*) is shown in Fig. 6.7.

The hierarchical structure of the created model is shown in the figure. This is a *FlatTank* model that calls the *LimitValue* function from the *Functions* package (Fig. 6.8).

After simulating the model for 350 s, we see that the filling level of the tank began to increase, reaching and then overcoming the height h_{ref}. As soon as the h_{ref} height was passed, the exit valve opened, and after 150 s the level stabilized. However, at this point, the input stream suddenly increased, raising the water level before the regulator managed to stabilize it again. This is displayed in Fig. 6.9.

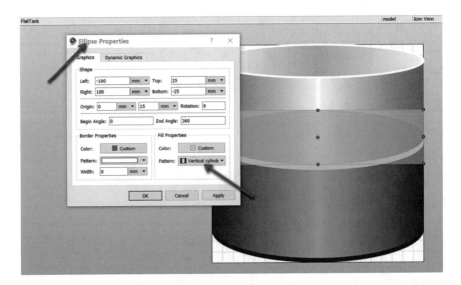

Fig. 6.7 Building a *FlatTank* icon (flat bottom tank)

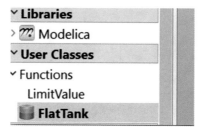

Fig. 6.8 Hierarchical structure of the created *FlatTank* model

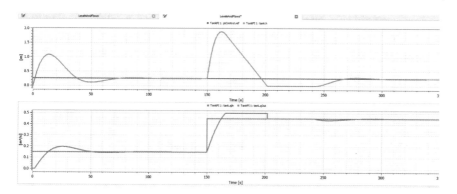

Fig. 6.9 Current liquid level in the tank (yellow graph) compared to the set (blue) is shown in the upper graph, with changing flows at the inlet (red graph) and at the outlet (green) are shown in the lower graph

Let us move on to component modeling and creating a full-fledged hierarchical structure. When using an object-oriented approach to modeling based on components, it is necessary to understand the structure of the system and the possibility of its hierarchical division into simpler objects from top to bottom. After the system of components and the interaction schemes between these components are defined, we can proceed to the first phase of modeling, identifying variables, and making equations for each component of the model.

In Fig. 6.4, five components can be distinguished:

- storage tank;
- fluid source;
- level sensor;
- valve;
- regulator.

The next step is to determine the ways of interaction and data transfer between components. Obviously, fluid flows from the source to the reservoir through the pipe. Fluid also leaves the reservoir through a valve-controlled outlet. The regulator

Fig. 6.10 Graphical representation of an object-oriented reservoir model

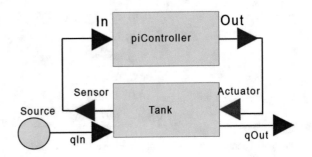

must receive fluid level measurements from the sensor. Thus, a data channel must be established between the sensor on the tank and the regulator.

As we already know from the previous sections, for this we need connectors.

And finally, it is convenient to create a library of your own components with the aim of reusing them in other models. Then, for example, we can connect various types of regulators to the finished "tank" component.

The structure of a reservoir model developed using an object-oriented approach based on components is shown in Fig. 6.10.

Three different types of classes can be distinguished that will be used in the model:

- *Interfaces*;
- *Functions*;
- *Components*.

Create a package containing three corresponding subpackages. To create a new package, right-click on the root directory of *User Classes* in the *Class Browser* window and select the *New Class* command, as shown in Fig. 6.11. You can also right-click on the package to which you want to add your package and select the *New Class* command.

In the dialog box that opens, select the type of class to create, in this case *Package*, and set the name *Hierarchical*. Click on the **OK** button to complete the creation of a new package. The package will appear in the hierarchical view of the *User Classes* directory in the *Class Browser* window, Fig. 6.12. With a right-click on the newly created package, we can create and add new models and packages to it.

Inside the created *Hierarchical* package, create a new subpacket where we will store the connectors of our model. Let us call it *Interfaces* (Fig. 6.13).

If it has not yet been expanded, open it in the *Class Browser* window to view its contents. Create the connector classes inside the *Interfaces* package, specifying Connector as the class type in the *New Class* dialog box (Fig. 6.14).

Each pair of related components requires input and output connectors. For example, the tank must be connected to a PI controller. To establish this connection, create two connectors (input and output) for reading the liquid level:

```
connector ReadSignalOutput = output Real(unit = "m")
"Output reading of tank level"
connector ReadSignalInput = input Real(unit = "m") "Input
```

Fig. 6.11 Menu for creating a new class

Fig. 6.12 Creating a new package for a hierarchical model

Fig. 6.13 Create a new subpackage for storing connectors

```
reading of tank level"
```

Let us switch to the icon creation mode (how to do this was described in detail in the second chapter) and create an icon in the form of a green triangle for entry and a white triangle for exit (Fig. 6.15).

In the future, when creating the components, we will place the white connector on the tank, and the green one on the PI controller and connect them.

Let us create two connectors for the reverse signal transmission from the PI controller to the tank for setting the valve position.

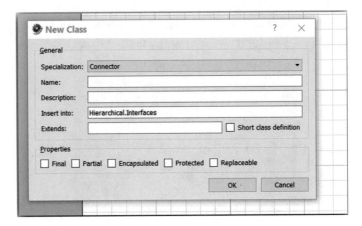

Fig. 6.14 Creating a connector in the hierarchical interfaces subpackage

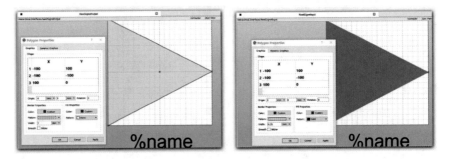

Fig. 6.15 Creation of pictograms of connectors for communication between the tank and the PI controller

```
connector ActSignalInput = input Real(unit = "m3/s")
"Input of signal to actuator for setting valve position"
connector ActSignalOutput = output Real(unit = "m3/s")
"Output of signal to actuator for setting valve position"
```

Create the pictograms again. Now use red color (Fig. 6.16):
And finally, create connectors that connect the fluid flow to the tank inlet:

```
connector LiquidFlowOutput = output Real(unit = "m3/s")
"Output of liquid flow at outlets"
connector LiquidFlowInput = input Real(unit = "m3/s")
"Input of liquid flow at inlets"
```

Create the pictograms again. Use blue color (Fig. 6.17).

A fluid stream flows from the reservoir, for it we use the same *LiquidFlowOutput* connector as for the inlet, but we will not connect it to anything.

Now, the hierarchical structure should look like this (Fig. 6.18).

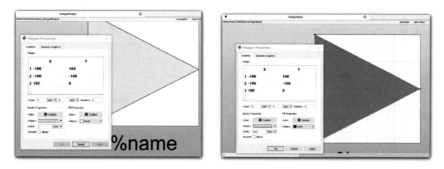

Fig. 6.16 Creation of pictograms of connectors for communication between PI controller and tank valve

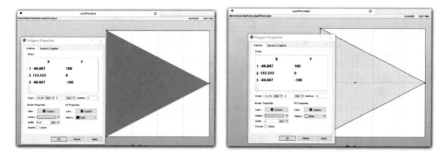

Fig. 6.17 Creation of pictograms of connectors for connecting the fluid flow with the inlet of the tank

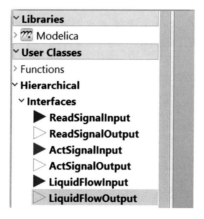

Fig. 6.18 Created subpackage of connectors

The next step is to create the three components of the system. Let us start by creating the *Components* subpackage inside the *Hierarchical* package. In the *Components* package, create a reservoir component called *Tank*. As mentioned above, the "reservoir" contains four connectors, we define the variables that are transmitted using them

- Connector LiquidFlowInput transmits a variable input stream *qIn*,
- Connector LiquidFlowOutput transmits a variable output flow *qOut*;
- Connector ReadSignalOutput transmits a fluid level variable *y*;
- Connector ActSignalInput transmits a signal from the PI controller *u*.

We write the equations that determine the behavior of the liquid in the tank. The central equation is the differential equation we derived above:

```
der(h) = (qIn - qOut) / A
```

The flow at the outlet of the tank is related to the position of the valve through the PI controller variable −*u* and the *LimitValue* function. When creating this model, we used the function *LimitValue* that we created in the previous approach to solving this problem. It is still stored in our *Functions* package.

A general text view of the reservoir component is shown in Fig. 6.19. For the geometric dimensions of the tank and the restrictions for the flow, the same numerical values were used as in the previous task.

Leave the text mode and go to *Diagram View*. We see only connectors—two input and two output (Fig. 6.20).

Now, we will go into the icon creation mode and create a conditional image of the tank in the same way as in the previous task. Connectors will automatically appear on the icon (Fig. 6.21).

Create a component of the *Flow* fluid source that generates the fluid flow. We also create it in the *Components* package:

```
model· LiquidSource;
    parameter Real flowLevel=0.15;
```

```
                                   Tank
Hierarchical.Components.Tank                                    model    Mode
model Tank "Model of a simple tank holding liquid"
    parameter Real A(unit = "m2") = 1.0 "Bottom area";
    parameter Real minFlow(unit = "m3/s") = 0.0 "Minimum flow through output valve";
    parameter Real maxFlow(unit = "m3/s") = 0.5 "Maximum flow through output valve";
    Interfaces.ActSignalInput u "Actuator controlling output flow, connector" ª;
    Interfaces.LiquidFlowInput qIn "Flow through input valve, connector" ª;
    Interfaces.ReadSignalOutput y "Sensor reading tank level, connector" ª;
    Interfaces.LiquidFlowOutput qOut "Flow through output valve, connector" ª;
    Real h(start = 0.0, unit = "m") "Liquid level";
    Real qOutMax(unit = "m3/s") "Maximum output flow considering that the tank level cannot be negative";
equation
    assert(minFlow >= 0, "minFlow - minimum flow through output valve must be >= 0");
    der(h) = (qIn - qOut) / A;
    qOutMax = if h > 0 then maxFlow else min(qIn, maxFlow);
    qOut = Functions.LimitValue(minFlow, qOutMax, -u);
    y = h;
    ª;
end Tank;
```

Fig. 6.19 Screen text view of the tank model (tank)

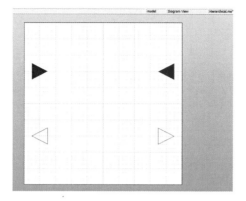

Fig. 6.20 Screen graphic view of the tank model (tank)

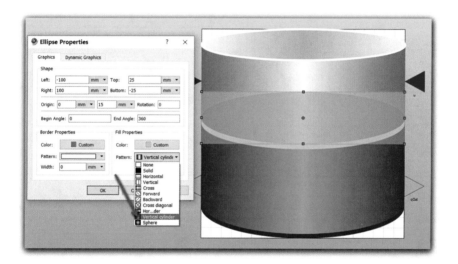

Fig. 6.21 Building the tank icon

```
    Hierarchical.Interfaces.LiquidFlow qOut;
equation
    qOut.lflow=if time > 150 then 3*flowLevel else flowLevel;
end LiquidSource;
```

In fact, it does not contain equations, but only the condition for the instantaneous increase in flow at a certain point in time.

In a graphical representation, it contains one connector, which we then connect to the tank, Fig. 6.22.

Now, we will go into the icon creation mode and create a conditional image of the liquid source. We draw the image as an *Ellipse* shape, following the pattern of a sphere. The selection of the appropriate shape and sample is shown in Fig. 6.23.

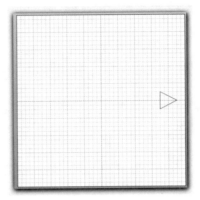

Fig. 6.22 On-screen graphical view of the *LiquidSource* model

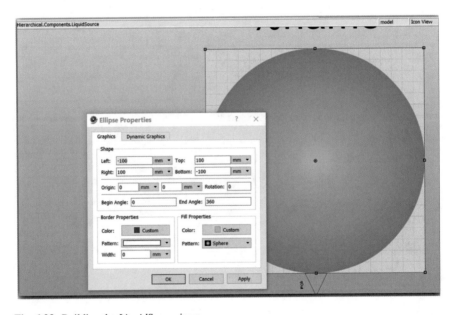

Fig. 6.23 Building the *LiquidSource* icon

Now the hierarchical structure should look like Fig. 6.24.

Define the regulators. We will build them on the basis of the "open" (incomplete) class of the *BaseController* controller that contains the main parameters, state variables, and two connectors: the first is for reading the liquid level sensor, the second is for controlling the tank valve.

```
partial model BaseController "Base class for tank level
controllers"
  parameter Real ref(unit = "m") "Reference level";
  Interfaces.ActSignalOutput u "Control to actuator,
```

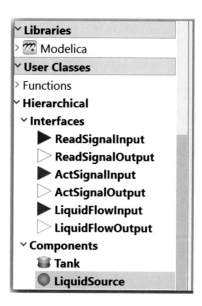

Fig. 6.24 Connector, component, and function structure

```
connector";
   Interfaces.ReadSignalInput y "Input sensor level,
connector";
   Real e(unit = "m") "Deviation from reference level";
equation
   e = ref - y;
end BaseController;
```

This incomplete model simply calculates the magnitude of the error e. Let us turn to the graphic view, Fig. 6.25.

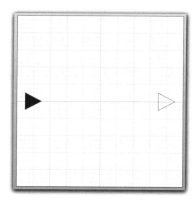

Fig. 6.25 *BaseController* screen graphic

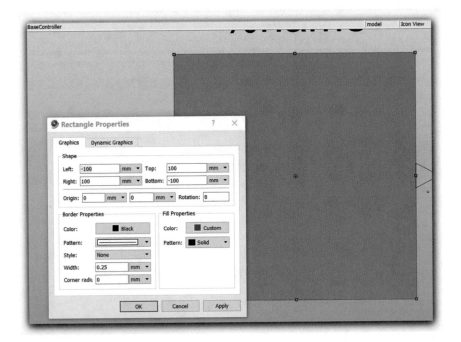

Fig. 6.26 Building the *BaseController* icon

Let us build an icon. We draw the image as a rectangle *Rectangle*, according to the pattern "solid". The selection of the appropriate shape and sample is shown in Fig. 6.26.

We proceed to the construction of the content of the regulators. First, we use the PI controller, which we will later replace with other types of regulators. We already know that the behavior of the PI controller (proportional-integral controller) is determined by the following equations:

$$\frac{dx(t)}{dt} = \frac{1}{T}e(t)$$
$$u_h(t) = K \cdot (e(t) + x(t))$$

We put these two equations in the PIController controller class.

```
model PIController "Elementary PI controller"
   extends BaseController;
   parameter Real K(unit = "m2/s") = 0.1 "Gain";
   parameter Real T(unit = "s") = 10 "Integrator time
constant";
   Real x(unit = "m") "Integrator state";
equation
   der(x) = e / T;
   u = K * (e + x);
```

```
end PIController;
```

The class (component) that we created is inherited; it inherits variables and equations from the parent class *BaseController*. Let us check the model for balance and create, as usual, an icon, Fig. 6.27.

We have created a complete hierarchical structure of our component model. Now, it should look like this (Fig. 6.28).

You can start compiling a system model by dragging and dropping components onto a diagram (Fig. 6.29).

You can go to the text view. We will see the program code of the *TankPI* model in Modelica, as shown in Fig. 6.30.

Create an icon using the ready-made drawing of the tank and adding the PI signature to the field in text mode, Fig. 6.31.

Fig. 6.27 PIController icon

Fig. 6.28 Structure of the component model of the tank with a PI controller

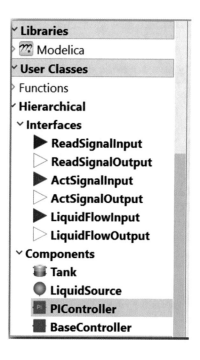

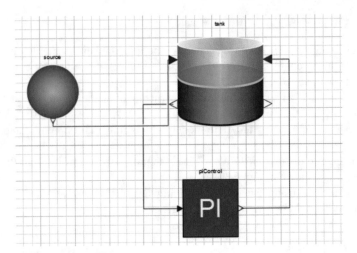

Fig. 6.29 On-screen graphic view of the model with one tank and PI controller

```
                                TankPI                                    ☒
Hierarchical.TankPI
model TankPI "A tank controlled by a PI controller"
  IntroductoryExamples.Hierarchical.Components.LiquidSource source ¤;
  IntroductoryExamples.Hierarchical.Components.PIController piControl(ref = 0.25) ¤;
  IntroductoryExamples.Hierarchical.Components.Tank tank ¤;
equation
  connect(source.qOut, tank.qIn) ¤;
  connect(tank.y, piControl.y) ¤;
  connect(piControl.u, tank.u) ¤;
  ¤;
end TankPI;
```

Fig. 6.30 Screen text view of the model with one tank PI controller

Performing a numerical experiment for 350 s gives exactly the same result as for a system with a flat bottom, which we built without component modeling (Fig. 6.32).

Now create a *TankPID* system identical to the *TankPI* system, except that the PI controller is replaced by the PID controller. Here, you can see the clear superiority of the object-oriented component approach over the traditional model approach, since the system components can be easily replaced and changed in a plug-and-play style. The proportional-integral-differential model (PID) of the controller can be performed in the same way as the model of the PI controller.

The basic equations of the PID controller are as follows:

$$e(t) = h_{\text{ref}} - h(t)$$

$$u_{\text{h}}(t) = K_{\text{p}} \cdot e(t) + \frac{1}{T_{\text{I}}} \cdot \int_0^t e(t) \cdot \mathrm{d}t + K_{\text{d}} \frac{\mathrm{d}e(t)}{\mathrm{d}t}$$

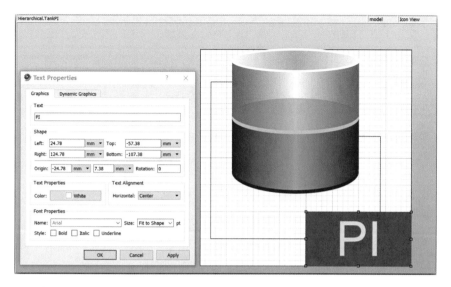

Fig. 6.31 Icon of model with one tank and PI controller

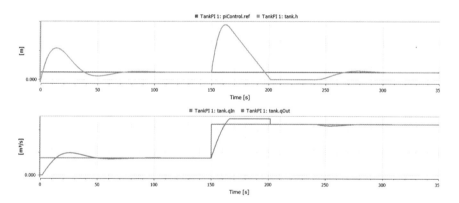

Fig. 6.32 Current liquid level in the tank (yellow graph) compared to the set (blue) is shown in the upper graph, with changing flows at the inlet (red graph) and at the outlet (green) are shown in the lower graph

where K_p is the gain of the proportional component (proportional to the PID controller parameter), T_I is the integration constant (integral parameter of the PID controller), and K_d—gain of the differential component. To create code in Modelica, we will get rid of the integral component. To do this, we introduce the notation for the error signal integral, and denote the derivative of the error signal as $y(t)$:

$$I(t) = \int_0^t e(t) \cdot dt$$

Then, the equations will take the following differential forms:

$$\frac{dI(t)}{dt} = e(t)$$

$$u_h(t) = K_p \cdot \left(e(t) + \frac{1}{K_p T_I} \cdot I(t) + \frac{K_d}{K_p} \frac{de(t)}{dt} \right)$$

We introduce the following convenient notation:

$$\frac{1}{K_p T_I} \cdot I(t) = x(t), \quad K_p T_I = T, \quad K_p = K, \quad \frac{K_d}{K_p} = T_d$$

Finally, the equations are rewritten as:

$$\frac{dx(t)}{dt} = \frac{1}{T} e(t)$$

$$u_h(t) = K \cdot \left(e(t) + x(t) + T_d \frac{de(t)}{dt} \right)$$

Here, $x(t)$ is the state variable of the PID controller, T is the time constant of the integral gain, K is the integral gain of the PID controller, and T_d is the time constant of the differential gain.

Using these equations and the parent class *BaseController*, we can create an inherited class *PIDController*

```
model PIDController "Elementary PID controller"
    extends BaseController;
    parameter Real K(unit = "m2/s") = 0.1 "Gain";
    parameter Real T(unit = "s") = 10 "Integrator time
constant";
    parameter Real Td(unit = "s") = 5 "Derivative gain";
    Real x(unit = "m") "Integrator state";
equation
    der(x) = e / T;
    u = K * (e + x + Td * der(e));
end PIDController;
```

Check the model for balance and create, as usual, an icon, Fig. 6.33.

You can begin compiling a system model by dragging and dropping components onto a diagram. However, it is easier to obtain a tank with a PID controller by replacing the PI controller with the PID controller in the previous diagram, Fig. 6.34.

You can go to the text view. We will see the program code of the TankPID model in Modelica:

```
model TankPID "A tank controlled by a PID controller"
    Components.Tank tank;
    Components.PIDController pidControl(ref = 0.25);
    Components.LiquidSource source;
equation
```

Fig. 6.33 PIDController icon

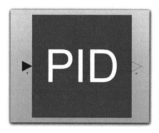

Fig. 6.34 On-screen graphic view of the model with one tank and PI controller

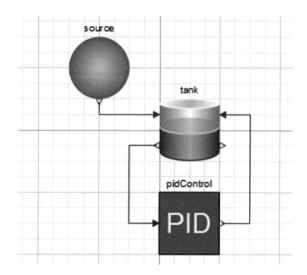

```
    connect(source.qOut, tank.qIn);
    connect(tank.y, pidControl.y);
    connect(pidControl.u, tank.u);
  end TankPID;
```

Let us create an icon using a ready-made picture of the tank and adding the PID signature in the text mode to the field, as shown in Fig. 6.35.

Run the constructed model for a numerical experiment.

In this model, the PI controller and PID controller gave us the same results; compare Figs. 6.32 and 6.36.

Modeling and computational Experiment 3

We will build a more complex system, taking advantage of the object-oriented component approach. We compose a system of three tanks, Fig. 6.37.

In textual form, the model looks like this:

```
model TankSystem "System of multiple connected tanks"
Components.PIController piControl1(ref = 0.2);
  Components.PIController piControl2(ref = 0.4;
```

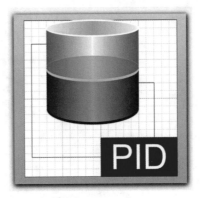

Fig. 6.35 Single tank model icon with PID controller

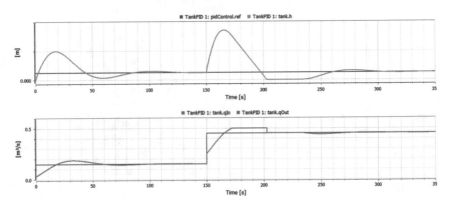

Fig. 6.36 Current liquid level in the tank (yellow graph) compared to the set (blue) is shown in the upper graph, with changing flows at the inlet (red graph) and at the outlet (green) are shown in the lower graph

```
    Components.PIController piControl3(ref = 0.2;
Components.Tank tank3;
    Components.Tank tank2;
    Components.Tank tank1;
Components.LiquidSource source;
equation
    connect(source.qOut, tank1.qIn);
    connect(piControl1.u, tank1.u);
    connect(tank1.y, piControl1.y);
    connect(tank1.qOut, tank2.qIn);
    connect(piControl2.u, tank2.u);
    connect(tank2.y, piControl2.y);
    connect(tank2.qOut, tank3.qIn);
    connect(piControl3.u, tank3.u);
    connect(tank3.y, piControl3.y);
end TankSystem;
```

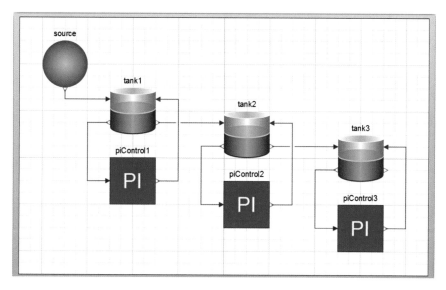

Fig. 6.37 System of three tanks. Component approach

When working with complex component systems, you should carefully consider the choice of system parameters and initial values. It should be noted that we can set parameter values at the bottom of the screen in the same way as we did with ready-made components from standard libraries. By clicking on the selected component, a window with the parameters available for change will open, as shown in Fig. 6.38.

In this model, we will set the required level of 0.2 m for the first and third tanks and 0.4 m for the second tank.

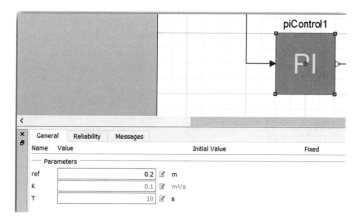

Fig. 6.38 Selection of the parameters of the first PI controller

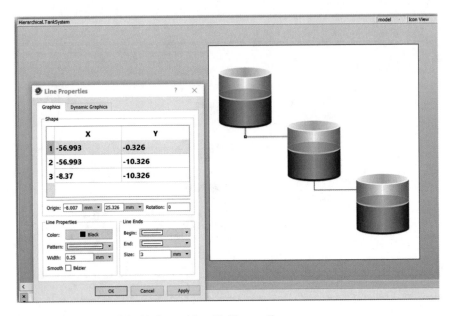

Fig. 6.39 Icon of a model with three tanks with PI controllers

We will prepare an icon. In this case, we still have to additionally use the Line form to connect the tanks together (Fig. 6.39).

Let us conduct a numerical experiment. We see that the stabilization of flows occurred by 150 s, but after increasing the input flow by a factor of three, the system again lost its equilibrium and the final stabilization occurred only by 1000 s (Fig. 6.40).

At the end, we present the hierarchical structure of all the tasks that we completed in this section, Fig. 6.41.

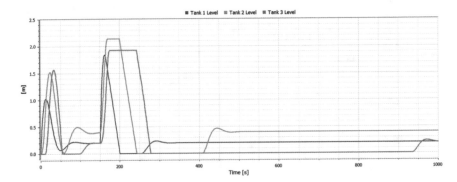

Fig. 6.40 Liquid level in the first, second, and third reservoir

Fig. 6.41 Hierarchical structure of the reservoir component model

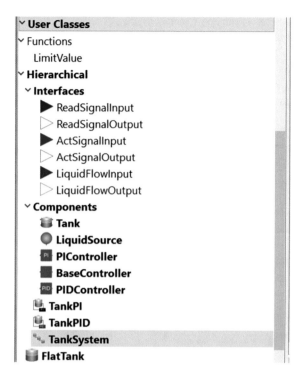

6.2 Heating a Liquid Mixture

Formulation of the problem

This problem is adopted from [4]. Figure 6.42 presents a system model consisting of two tanks, a fluid source, a pump, and a heating element. Tank 1 and Tank 2 have a capacity of 10 m^3 and 1 m^3, respectively. The reservoir 1 is connected to a fluid

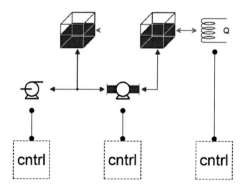

Fig. 6.42 Graphic model of a tank system

source. Three controllers are provided for controlling the source, pump, and heater (*cntrl*).

In the problem under consideration, a liquid is a mixture consisting of 25% kerosene and 75% gasoline in a percentage ratio. The density of kerosene and gasoline is 760 kg/m³ and 849 kg/m³, respectively. The heat capacities of kerosene and gasoline in the temperature range of interest to us can be approximated by the following linear functions: for kerosene $C_p = 446 + 5.36T$ and for gasoline $C_p = 325 + 4.60T$, J/(kg K).

The rate of outflow or outflow of fluid from the source, supported by the first regulator is $Q_{mass1} = 40$ kg/s. The pump pumps liquid between the tanks at a speed supported by the second regulator, equal to $Q_{mass2} = 10$ kg/s. The temperature of the liquid at the initial time $T = 300$ K. The reservoir 2 is heated using a heating element. The heat flow rate of the heater supported by the third regulator $Q_{heat} = 2.5 \times 10^5$ J/s.

Since the reservoir is closed, the volume of fluid stored in the reservoir cannot be greater than its volume.

When controlling the source and pump, two extreme situations must be considered:

(1)   the inability to extract fluid from an empty tank;
(2)   the inability to add fluid to a completely filled tank.

The tank is supposed to be empty if the mass of liquid in it does not exceed 1 kg.

Tasks

Build a hierarchical component model of the next dynamic system. Two tanks are initially empty. The operation of the system is defined by the following sequence:

1.   From the initial moment of time $t = 0$ to $t_1 = 180$ s, the source with the liquid supplies 1 solution to the tank;
2.   At time $t_2 = 240$ s, the pump turns on and starts pumping liquid from Tank 1 to Tank 2, and works for 2 min. All that is needed for this is a completely liquid;
3.   At time $t_3 = 360$ s, the heater is turned on, the heated fluid located in the second tank. The heater runs for 6 min $t_4 = 720$ s;
4.   At $t_4 = 720$ s, the pump switches on again and starts pumping liquid from the second tank to the first. The regulator must maintain a constant flow rate for 3 min from $t_4 = 720$ s to $t_5 = 900$ s. However, in the second part of the tank, it is completely empty; further pumping of liquid is impossible;
5.   When starting from $t_6 = 840$ s, the liquid flowing from the first tank back to the source, work programs end at $t_9 = 1200$ s.

Independently conduct modeling, simulation, and analysis of this task in the WSM environment, observing the hierarchical architecture, as shown in Fig. 6.43.

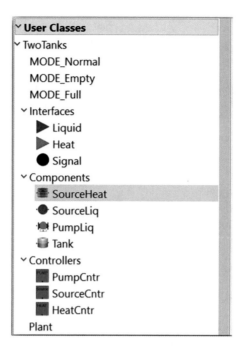

Fig. 6.43 Hierarchical architecture of the model of two tanks

The library (package) you created named *TwoTanks* should consist of the following subpackages:

- *Interfaces* package containing connectors for transferring fluid, heat, and control (in Fig. 6.43 they are represented by a blue, red triangle, and black dot, respectively);
- *Components* package, which contains a description of the classes required for a complete model: (*SourceLig*) fluid source, (*SourceHeat*) heater, (*PumpLig*) pump, and (*Tank*) tank;
- *Controllers* package, which contains a description of the three control knobs;
- *Plant* class describing the entire system.

Modeling and computational experiment

The process of creating components in textual, graphical, and pictorial representations was examined in detail in the first example. Therefore, now briefly dwell on the mathematical foundations of creating each component and the principles of constructing a common model.

Let us start by creating a package of regulators.

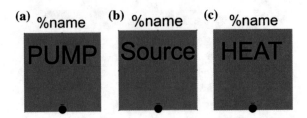

Fig. 6.44 Pictograms of regulators, **a** *PumpCntr* pump, **b** *SourceCntrq* fluid source, and **c** *HeatCntr* heater

Regulators (Fig. 6.44).

Pump regulator *PumpCntr* must maintain a constant rate of fluid transfer from one tank to another and vice versa $Q_{mass2} = 10$ kg/s. To do this, we introduce two equal in magnitude parameter of the fluid flow

```
parameter SI.MassFlowRate flow1 = 10 "direct fluid flow"
parameter SI.MassFlowRate flow2 = 10 "reverse fluid flow"
```

In accordance with the condition of the problem, we will create four events: the beginning and completion of pumping fluid from the first tank to the second and, similarly, the beginning and completion of pumping fluid from the second tank to the first:

```
parameter SI.Time event1 = 240, "start of direct fluid transfer"
                   event2 = 360, "completion of direct pumping"
                   event3 = 720, "start of fluid transfer"
                   event4 = 900; "completion of fluid transfer"
```

In the intervals between these events, various fixed values of the *totalMassFSP* flow are set, which will be transmitted as an input signal to the pump (see the equation block of equations in the program code)

```
signal.s = {totalMassFSP};
```

Pump controller program code:

```
model PumpCntr
Interfaces.Signal signal(nSignal=1)
parameter SI.Time event1 = 240,
event2 = 360,
event3 = 720,
event4 = 900;
parameter SI.MassFlowRate flow1 = 10,
flow2 = 10;
SI.MassFlowRate totalMassFSP;
equation
totalMassFSP = if time > event4 then 0
else if time > event3 then -flow1
else if time > event2 then 0
else if time > event1 then flow2
else 0;
signal.s = {totalMassFSP};
end PumpCntr;
```

Fluid source regulator *SourceCntrq* has to maintain a constant rate of fluid transfer from the source to the tank and back $Q_{mass1} = 40$ kg/s. To do this, we introduce two equal in magnitude parameter of the fluid flow:

```
parameter SI.MassFlowRate flow1 = 40, "direct fluid flow"
                          flow2 = 40; "reverse fluid flow"
```

In accordance with the condition of the problem, we create two events: the end of the fluid supply to the tank and, similarly, the beginning of the flow of fluid from the tank:

```
parameter SI.Time event1 = 180, "completion of direct pumping"
                  event2 = 840; "start of fluid transfer"
```

In the intervals between these events, various fixed values of the *totalMassFSP* flow are set, which will be transmitted as an input signal to the pump (see the `equation` block of equations in the program code). In addition, we must supply a mixture of fluids of a given temperature. We will also control these parameters. Therefore, the input to the fluid source is a vector quantity

```
signal.s[1:nComp] = massFractSP; // the proportion of substances in the mixture
signal.s[nComp+1] = totalMassFSP; // fluid flow rate
signal.s[nComp+2] = tempFSP; // liquid temperature (mixture)
```

Liquid Source Regulator Program Code:

```
model SourceCntr
parameter Integer nComp = 2;
Interfaces.Signal signal(nSignal=nComp+2)
parameter SI.Time event1 = 180,
event2 = 840;
parameter SI.MassFlowRate flow1 = 40,
flow2 = 40;
SI.MassFlowRate totalMassFSP;
Real massFractSP[nComp];
SI.Temperature tempFSP;
equation
totalMassFSP = if time > event2 then flow2
else if time > event1 then 0
else -flow1;
massFractSP = {0.25, 0.75};
tempFSP = 300;
signal.s[1:nComp] = massFractSP;
signal.s[nComp+1] = totalMassFSP;
signal.s[nComp+2] = tempFSP;
end SourceCntr;
```

Heater regulator *HeatCntr* must maintain a constant flow of heat to heat the fluid $Q_{heat} = 2.5\ 10^5$ J/s. To do this, we introduce the heat flux parameter:

```
parameter SI.EnergyFlowRate heatFlowSP = 2.5E5; "heating value"
```

In accordance with the condition of the problem, we create two events: the beginning and end of heating the liquid in the tank:

```
parameter SI.Time event1 = 360, "start heating fluid"
                  event2 = 720; "completion of fluid heating"
```

In the intervals between these events, the liquid is heated (see the `equation` block in the program code) using a fixed heat flux `heatFSP`, the value of which is transmitted to the heater

```
signal.s = {heatFSP};
```

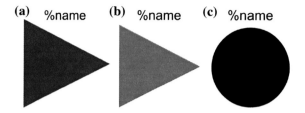

Fig. 6.45 Connector icons for **a** liquid, **b** heating, and **c** control

Heater Controller Program Code:

```
model HeatCntr
Interfaces.Signal signal(nSignal=1)
parameter SI.EnergyFlowRate heatFlowSP = 2.5E5;
parameter SI.Time event1 = 360,
event2 = 720;
SI.EnergyFlowRate heatFSP;
equation
heatFSP = if time > event1 and time < event2
then -heatFlowSP
else 0;
signal.s = {heatFSP};
end HeatCntr;
```

Now create a connector package.

Connectors

The blue connector is designed to connect the reservoir to the fluid source and to the pump, taking into account the current fluid temperature (Fig. 6.45):

```
connector Liquid
parameter Integer nComp=1;
Integer mode;
SI.Mass mass[nComp] "Mass component of liquid kerosene and gasoline";
SI.Temperature temp "Temperature";
flow SI.MassFlowRate Fm[nComp] "fluid flow";
flow SI.EnergyFlowRate Fh "fluid transported heat";
end Liquid;
```

The red connector is for connecting the heater to the tank:

```
connector Heat
Integer mode;
flow SI.EnergyFlowRate Q; "heat flow from the heater"
end Heat;
```

The black connector is designed to transmit a signal from the controller to any component:

```
connector Signal
parameter Integer nSignal = 1;
Real s[nSignal]; "transmitted signal"
end Signal;
```

Fig. 6.46 *SourceLiq* fluid
source icon

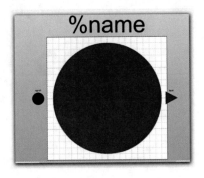

Now create the main component package.

Component: Fluid Source

Let us start by creating a *SourceLiq* fluid source. This component is connected with
the blue connector (*liq* variable) to the tank and with the black connector (*signal*
vector variable) with the regulator (Fig. 6.46).

Let us consider three tank states (these states will be described in detail in the
Tank component)

- mode $= M_{\mathrm{Empty}}$, liquid out of the tank is not possible;
- mode $= M_{\mathrm{Normal}}$, direct and reverse fluid flow possible;
- mode $= M_{\mathrm{Full}}$, no further filling of the tank.

The source behavior can be described by the following equations. When liq-
uid flows into the tank $Q_{\mathrm{mass1}} > 0$, the flow is $F_{\mathrm{m}} = 0$ if the tank is full
mode $= M_{\mathrm{Full}}$. Otherwise, the flow is calculated as the transfer of the mixture of
kerosene $signal.n(k)$ and gasoline $signal.n(b)$ specified by the regulator with the
speed specified by the regulator $signal.totalMass$

$$F_{\mathrm{m}}(k) = signal.n(k) \cdot signal.totalMass$$
$$F_{\mathrm{m}}(b) = signal.n(b) \cdot signal.totalMass$$

The temperature is determined by the regulator signal:

$$TempF = signal.TempF$$

When the fluid flows $Q_{\mathrm{mass1}} < 0$, the flow $F_{\mathrm{m}} = 0$, if the tank is empty mode $=$
M_{Empty}. Otherwise, the flow is calculated based on the current state of the liquid in
the tank

$$F_{\mathrm{m}}(k) = \frac{signal.totalMass \cdot liq.m(k)}{liq.m(k) + liq.m(b)}$$
$$F_{\mathrm{m}}(b) = \frac{signal.totalMass \cdot liq.m(b)}{liq.m(k) + liq.m(b)}$$

The temperature matches the current temperature in the tank

$$TempF = liq.T$$

Knowing in each case the temperature of the liquid *TempF*, we can record its heat capacity as a linear function of temperature:

$$C_p = C_{p0} + C_{p1} \cdot TempF$$

And the amount of heat, respectively, received or given away by the liquid

$$F_h = F_m \cdot C_p \cdot TempF$$

Liquid Source Program Code:

```
model SourceLiq
parameter Integer nComp = 1;
Interfaces.Liquid liq(nComp=nComp)
Interfaces.Signal signal(nSignal=nComp+2)
// Signals from the regulator
Real sFracMass[nComp];
SI.MassFlowRate sTotalMassF;
SI.Temperature sTempF;
// Specific heat
parameter Real CpCoefs[nComp,2];
protected
SI.SpecificHeatCapacity Cp[nComp];
SI.Temperature TempF;
equation
signal.s[1:nComp] = sFracMass;
signal.s[nComp+1] = sTotalMassF;
signal.s[nComp+2] = sTempF;
if sTotalMassF > 0 then
// Fluid flow from reservoir to source
liq.Fm = if noEvent(liq.mode == MODE_Empty)
then zeros(nComp)
else sTotalMassF*liq.mass/sum(liq.mass);
TempF = liq.temp;
else
// Fluid flow from source to reservoir
liq.Fm = if noEvent(liq.mode == MODE_Full)
then zeros(nComp)
else sTotalMassF*sFracMass;
TempF = sTempF;
end if;
for i in 1:nComp loop
Cp[i] = CpCoefs[i,1] + CpCoefs[i,2]*TempF;
end for;
liq.Fh = Cp*TempF*liq.Fm;
end SourceLiq;
```

Component: Heater

Create a heater *SourceHeat* (Fig. 6.47).

This component is responsible for heating the second tank and is connected with the red connector to the tank (variable *heat*), and the black connector with the regulator (variable *signal*).

The behavior of the heater is completely determined by the regulator. If the tank is empty mode $= M_{\text{Empty}}$, then the heat flux $Q = 0$, otherwise $Q = signal$.
Heater Program Code:

```
model SourceHeat
Interfaces.Heat heat
Interfaces.Signal signal(nSignal=1)
equation
{heat.Q} = if noEvent(heat.mode == MODE_Empty)
then {0} else signal.s;
end SourceHeat;
```

Component: Pump

Create a pump *PumpLiq* (Fig. 6.48).

This component is responsible for transferring fluid from one tank to another and is connected by blue connectors to the tanks (*liqIn* variables—the fluid that flows

Fig. 6.47 *SourceHeat* heater icon

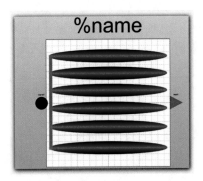

Fig. 6.48 *PumpLiq* pump icon

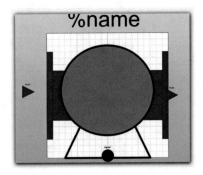

into the pump—the input variable, *liqOut*—the fluid that flows from the pump—the output variable), and the black connector—with the regulator (*signal* variable).

The pump behavior can be described by the following equations. When pumping fluid from the first to the second tank, $Q_{mass2} > 0$, the flow is zero $F_m = 0$, if the first tank is empty mode $= M_{Empty}$ or the second tank is full mode $= M_{Full}$. Otherwise, the flow is calculated as the transfer of mixture fractions with the speed specified by the regulator *signal.totalMass*

$$F_m(k) = \frac{signal.totalMass \cdot liqIn.m(k)}{liqIn.m(k) + liqIn.m(b)}$$

$$F_m(b) = \frac{signal.totalMass \cdot liqIn.m(b)}{liqIn.m(k) + liqIn.m(b)}$$

The temperature matches the current temperature in the first tank

$$TempF = liqIn.T$$

When pumping fluid from the second to the first reservoir, $Q_{mass2} < 0$, the flow is zero $F_m = 0$, if the second reservoir is empty mode $= M_{Empty}$ or the first reservoir is full mode $= M_{Full}$. Otherwise, the flow is calculated as the transfer of mixture fractions with the speed specified by the regulator *signal.totalMass*

$$F_m(k) = \frac{signal.totalMass \cdot liqOut.m(k)}{liqOut.m(k) + liqOut.m(b)}$$

$$F_m(b) = \frac{signal.totalMass \cdot liqOut.m(b)}{liqOut.m(k) + liqOut.m(b)}$$

The temperature coincides with the current temperature in the second tank

$$TempF = liqOut.T$$

Knowing in each case the temperature of the liquid *TempF*, we can record its heat capacity as a linear function of temperature:

$$C_p = C_{p0} + C_{p1} \cdot TempF$$

And the amount of heat, respectively, received or given away by the liquid

$$F_h = F_m \cdot C_p \cdot TempF$$

Obviously, if we turn on the pump in the opposite direction, then

$$F_m Out = -F_m In$$
$$F_h Out = -F_h In$$

Pump Program Code:

```
model PumpLiq
parameter Integer nComp = 1;
parameter SI.Mass epsMass = 1;
Interfaces.Signal signal(nSignal=1)
Interfaces.Liquid liqIn(nComp=nComp)
Interfaces.Liquid liqOut(nComp=nComp)
// Perfect heat capacity
parameter Real CpCoefs[nComp,2];
// Control action
SI.MassFlowRate sTotalMassF;
protected
SI.SpecificHeatCapacity Cp[nComp];
SI.Temperature TempF;
equation
signal.s = {sTotalMassF};
if sTotalMassF > 0 then
// Fluid flow from liqIn to liqOut
liqIn.Fm = if noEvent( liqIn.mode == MODE_Empty or
liqOut.mode == MODE_Full )
then zeros(nComp)
else sTotalMassF*liqIn.mass/sum(liqIn.mass);
TempF = liqIn.temp;
else
// Fluid flow from liqIn to liqOut
liqIn.Fm = if noEvent( liqOut.mode == MODE_Empty or
liqIn.mode == MODE_Full )
then zeros(nComp)
else sTotalMassF*liqOut.mass/sum(liqOut.mass);
TempF = liqOut.temp;
end if;
liqOut.Fm = -liqIn.Fm;
for i in 1:nComp loop
Cp[i] = CpCoefs[i,1] + CpCoefs[i,2]*TempF;
end for;
liqIn.Fh = Cp*TempF*liqIn.Fm;
liqOut.Fh = -liqIn.Fh;
end PumpLiq;
```

Component: Tank

Create a *Tank* (Fig. 6.49).
This component is a reservoir for storing liquid and is connected by a blue connector to a source or pump (*liq* variable), and a red connector, if necessary, with a heater (*heat* variable). To describe the processes in the tank, it is necessary to set as parameters the specific heat and density of the substances in the mixture, the volume of the tank, the maximum mass of the liquid to which ε_m flows from the tank. The temperature variable *temp* initially assumes a value of 300 K.

First of all, we describe the limiting cases. Let *mass* be the mass, and V the current volume of fluid in the tank. Then three tank conditions are possible

%name

Fig. 6.49 Tank icon

- mode $= M_{\text{Empty}}$, the outflow of fluid from the reservoir is impossible, this occurs if the mass of fluid in the reservoir is less than the limit mass $\leq \varepsilon_m = 1$ кг;
- mode $= M_{\text{Normal}}$, if mass $> \varepsilon_m$, $VolumeTank > V$ direct and reverse fluid flow possible;
- mode $= M_{\text{Full}}$, further filling of the tank is not possible, this occurs if the volume of liquid begins to exceed the volume of the tank $VolumeTank \leq V$.

The processes in the tank are described by the equations:
Equation for fluid flow

$$\frac{dm}{dt} = F_m$$

The volume of liquid (mixture)

$$V = \frac{m(k)}{\rho_k} + \frac{m(b)}{\rho_b}$$

Knowing the temperature of the liquid *temp*, we can write its heat capacity as a linear function of temperature:

$$C_p = C_{p0} + C_{p1} \cdot temp$$

Heat transfer equation (F_h is the amount of heat carried by the fluid through the pump, Q is the amount of heat received from the heater)

$$\frac{dH}{dt} = F_h + Q$$
$$H = m \cdot C_p \cdot temp$$

Tank Program Code:

```
model Tank
parameter Integer nComp = 1;
Interfaces.Liquid liq(nComp=nComp)
Interfaces.Heat heat;
// Specific heat of substances in the mixture
parameter Real CpCoefs[nComp,2];
// The density of substances in the mixture
parameter SI.Density density[nComp];
SI.Mass mass[nComp] (start=epsMass*ones(nComp)/nComp, fixed=true);
SI.Temperature temp(start=300, fixed=true);
SI.Enthalpy enthalpy;
SI.SpecificHeatCapacity Cp[nComp];
parameter SI.Volume volumeTank = 1 "Tank volume";
SI.Volume volL "Fluid volume";
parameter SI.Mass epsMass = 1;
equation
// tank operating modes
liq.mode = heat.mode;
liq.mode = if not sum(mass) > epsMass then MODE_Empty
else if not volL < volumeTank then MODE_Full
else MODE_Normal;
// equations for fluid mass:
der(mass) = liq.Fm;
liq.mass = mass;
volL = sum( mass[i]/density[i] for i in 1:nComp);
// For energy:
for i in 1:nComp loop
Cp[i] = CpCoefs[i,1] + CpCoefs[i,2]*temp;
end for;
der(enthalpy) = liq.Fh + heat.Q;
enthalpy = liq.mass*Cp*temp;
liq.temp = temp;
end Tank;
```

Next, you need to combine the prepared components into one module. Let us draw a block diagram of our hierarchical model (Fig. 6.50):

In each created component, it is necessary to enter the parameters in accordance with Table 6.1.

Let us assemble the component model from the created components (Fig. 6.51):

We give the main results of a numerical experiment (Figs. 6.52, 6.53, 6.54, 6.55, 6.56, 6.57, and 6.58).

6.3 Inverted Pendulum Problem

Formulation of the problem

This problem is adopted from [5]. A dynamic mechanical system consists of an inverted (reverse) pendulum mounted on a motorized trolley through a hinge, without

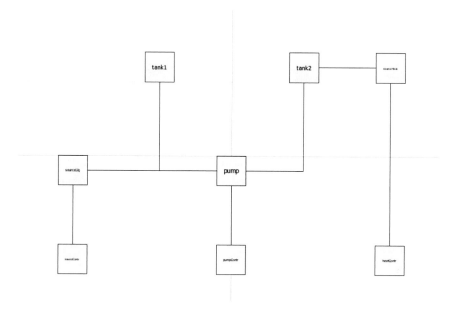

Fig. 6.50 Block diagram of a system of two tanks with heating

taking into account friction. The proposed pendulum system is an example commonly found in textbooks on control systems and other research literature. Attention to this problem is due to the constant instability of this system, namely the behavior of the pendulum, since the pendulum will simply fall if the trolley does not move, to keep it in equilibrium. In addition, the problem is interesting by the nonlinearity of the dynamics of the system. The task of the control system is to balance the inverted pendulum by applying force to the trolley to which the pendulum is attached. An example from real life that is directly related to this system is the control of the launch vehicle during take-off.

Consider a two-dimensional problem in which the pendulum is forced to move in a virtual vertical plane, shown in Fig. 6.59. For such a system, the control variable is the force f, which moves the trolley strictly in the horizontal plane, and the controlled variable is φ—the angular position of the pendulum and the horizontal position of the trolley.

Tasks

Build a hierarchical model of the inverse pendulum system.

Modeling and computational experiment

The basics of creating a mathematical model for such a system were described in the third chapter on the example of an elliptical pendulum. The main difference between an inverted pendulum from an elliptic is its instability. However, the main characteristics of the movement—the presence of portable acceleration in the pendulum,

Table 6.1 Parameters and constants used in the model

Title	Denotation	Value
The volume of the first tank	Vol1	10 m^3
The volume of the second tank	Vol2	1 m^3
The initial temperature of the liquid (mixture)	T	300 K
Kerosene density, ρ_k	rho[1]	760 kg/m^3
Gas density, ρ_b	rho[2]	849 kg/m^3
Kerosene heat capacity, zero-order coefficient C_{p0}	CpCoefs[1]	446 J/(kg K)
Kerosene heat capacity, first-order coefficient C_{p1}	CpCoefs[1, 2]	5.36 J/(kg K^2)
Heat capacity of gasoline, zero-order coefficient C_{p0}	CpCoefs[1, 2]	32 J/(kg K)
Heat capacity of gasoline, first-order coefficient C_{p1}	CpCoefs[2]	4.60 K/(kg K^2)
The time during which fluid flows into the first reservoir	t_1	0–180 s
The time during which the pump pumps fluid from the first reservoir to the second	t_2	240–360 s
Heater start time	t_3	360 s
Heater shutdown time	t_4	720 s
The time during which the pump pumps fluid from the second tank to the first	t_5	720–900 s
The time during which fluid flows from the tank	t_6	900–1200 s
The proportion of kerosene from the total mass of liquid	n_1	25%
The proportion of gasoline in total liquid	n_2	75%
The rate of fluid flow from the source to the first tank and vice versa	Q_{mass1}	40 kg/s
Pumping rate of liquid from the first tank to the second and vice versa	Q_{mass2}	10 kg/s
Heater heat flow	Q_{heat}	2.5×10^5 J/s

which arises as a result of its motion in a non-inertial reference system associated with the rectilinear uniformly accelerated movement of the trolley and the oscillatory motion of the pendulum—remain the same as in the elliptical.

The "inverted pendulum" system has two degrees of freedom and is located in the field of gravity. Let $q_1 = x$ be the generalized coordinates of the cart from the origin, and $q_2 = \varphi$ is the angle of deviation of the bar from the vertical as generalized coordinates, Fig. 6.59.

The Lagrangian of the inverted pendulum system is similar to the Lagrangian of an elliptical pendulum except that the sign of the angular position is measured from the vertical position of the unstable equilibrium, which leads to a change in the sign of the term containing the trigonometric function (see Sect. 3.3.5):

$$L = \frac{1}{2}(M + m)\dot{x}^2 + \frac{1}{2}ml^2\dot{\varphi}^2 - ml(\dot{\varphi}\dot{y} + g)\cos\varphi$$

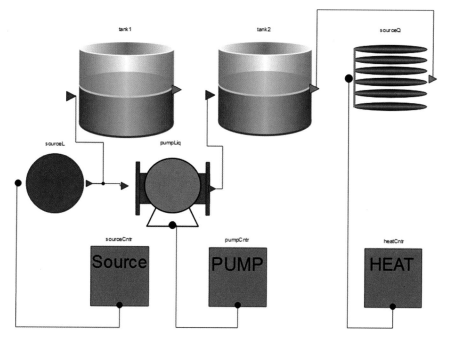

Fig. 6.51 Component hierarchical model "heating of two tanks"

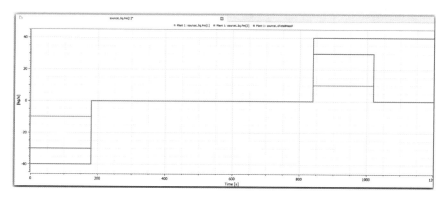

Fig. 6.52 Change in the fluid flow from the source in the first tank (red—total, for components of the gasoline and kerosene liquid—blue and green) over time

We take into account the friction and damping forces of the system using the Rayleigh dissipative function:

$$R = \frac{1}{2}b\dot{x}^2$$

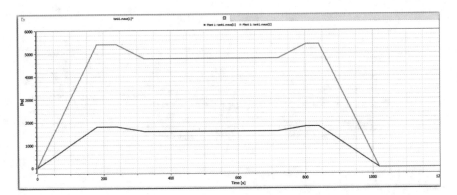

Fig. 6.53 Change in mass of the fluid components in the first tank over time (red for kerosene, blue for gasoline)

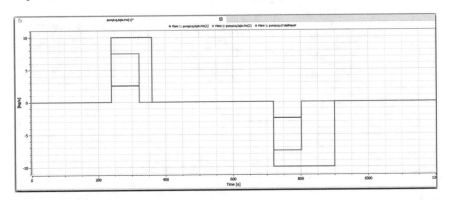

Fig. 6.54 Change in the fluid flow from the pump in the second tank (red—common, for the components of the benzene and kerosene liquid—blue and green) over time

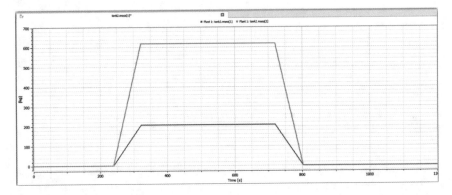

Fig. 6.55 Change in mass of the fluid components in the second tank over time (red for kerosene, blue for gasoline)

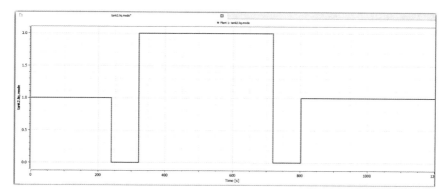

Fig. 6.56 Changing the state of the second tank in time

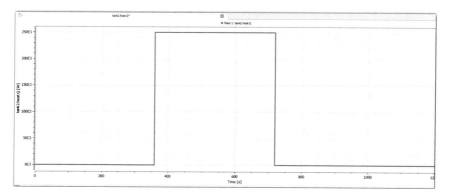

Fig. 6.57 Change in heat flux from the heater over time

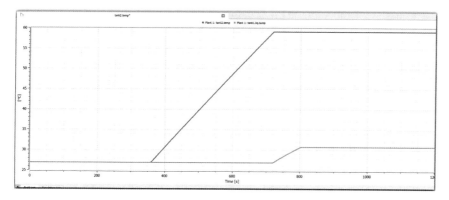

Fig. 6.58 Fluid temperature over time (blue graph for reservoir 2, yellow for reservoir 1)

Fig. 6.59 Physical model of
an inverted pendulum

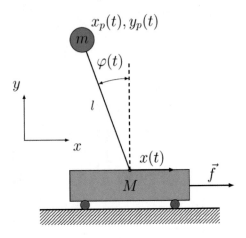

As a result, a term appears corresponding to the generalized force:

$$Q_{ext} = -\frac{\partial R}{\partial \dot{x}}$$

And finally, we take into account the external control action f. Performing transformations similar to those that we did to construct the model of an elliptic pendulum, we obtain a mathematical model of an inverted pendulum

$$(M + m)\ddot{x} - ml\ddot{\varphi}\cos\varphi + ml\dot{\varphi}^2\sin\varphi + b\dot{x} = f$$
$$l\ddot{\varphi} - g\sin\varphi - \ddot{x}\cos\varphi = 0$$

The basics of modeling a PID controller were described in Example 6.1. We already know that the output signal of the controller u is determined by three terms:

$$u(t) = P + I + D = K_p e(t) + K_i \int_0^t e(\tau)d\tau + K_d \frac{de}{dt}$$

where K_p, K_i, K_d are the gain factors of the proportional, integrating, and differentiating components of the controller, respectively. We assume that the vertical position of the pendulum is $\varphi_0 = 0$, and for the error, we take the deviation of the pendulum from the vertical $e(t) = (\varphi_0 - \varphi)$.

For the PID controller, we obtain the following system of equations:

$$e(t) = (\varphi_0 - \varphi)$$
$$u(t) = K_p P + K_i I + K_d D$$
$$P = e(t)$$

Fig. 6.60   Block diagram of a controlled inverted pendulum system

$$\frac{\mathrm{d}I}{\mathrm{d}x} = e(t)$$

$$D = \frac{\mathrm{d}e(t)}{\mathrm{d}t}$$

Let us construct the implementation of this task in the Wolfram SystemModeler mathematical modeling package, taking the output signal u of the PID controller for the control force f applied to the trolley.

We will create the *PIDReversPendulum* package, inside which we will create the *Pendulum* components, the PID controller and the *ControlRevPendulum* component that describes the entire system. The diagram of our hierarchical model will look like in Fig. 6.60.

The *PIDReversPendulum* package will appear in the hierarchical representation of the *User Classes* directory in the *Class Browser* window, and if everything is done correctly, it will look like Fig. 6.61.

In this example, we will not waste time creating our own connectors, but we will use the ready-made library *Modelica.Blocks.Interfaces*. We need two types of connectors—input and output (Fig. 6.62).

```
Modelica.Blocks.Interfaces.RealInput
Modelica.Blocks.Interfaces.RealOutput
```

In addition, for animating the control of an inverted pendulum, we will use the standard package of Visualizers of the form `Modelica.Mechanics.MultiBody.Visualizers.Advanced.Shape`. By choosing *Shape* with the right mouse button or simply by hovering over the

Fig. 6.61   Hierarchical structure of a controlled inverted pendulum system

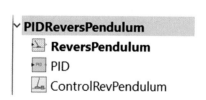

Fig. 6.62   Optional standard connector package

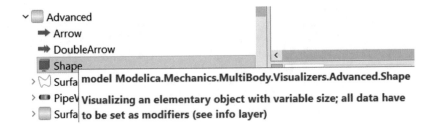

Fig. 6.63 Optional standard visualizer package

Fig. 6.64 Pictogram
"inverted pendulum on a
cart"

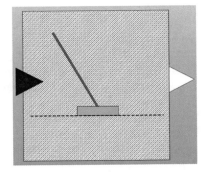

cursor, you can read additional information about creating the desired shape, as
shown in Fig. 6.63.

Let us move on to creating our own components.

Component: Inverted Pendulum on Trolley

Create the component "Inverted pendulum on a cart" (Fig. 6.64).

Using the *equations* we obtained and connecting the connector and visualizer
libraries, we can write the program code of the inverted pendulum.

```
model ReversPendulum
    parameter Modelica.SIunits.Length l = 0.3 "thread length";
    parameter Modelica.SIunits.Mass m = 0.2 "mass of the body";
    parameter Modelica.SIunits.Mass M = 0.5 "mass of the cart";
    parameter Real b(unit = "H.s2/m") = 0.19 "coefficient of
viscousity";
    parameter Real rad(unit = "m") = 0.05 "radius of the body";
    constant Real pi = 3.1416 "Pi";
    constant Real g(unit = "m/s2") = 9.81;
    Real x, xp, yp;
    Modelica.Mechanics.MultiBody.Visualizers.Advanced.Shape
    thread (shapeType = "cylinder", length = 2 * l, width = 0.01,
    height = 0.01, r = {x, 0, 0},
    lengthDirection = {xp - x, yp, 0}, color = {0, 50, 255});
    Modelica.Mechanics.MultiBody.Visualizers.Advanced.Shape
    bob(shapeType = "sphere", length = 2 * rad, width = 2 * rad,
```

```
  height = 2 * rad, r = {xp, yp, 0}, r_shape = {-rad,
  0, 0}, color = {255, 0, 0});
    Modelica.Mechanics.MultiBody.Visualizers.Advanced.Shape
  ground(shapeType = "box", length = 0.01, width = 2 * 1,
  height = 2 * 1, r = {0, -2 * 1, 0}, lengthDirection = {0,
  1, 0}, color = {0, 255, 0});
    Modelica.Mechanics.MultiBody.Visualizers.Advanced.Shape
  cart(shapeType = "box", length = 1 / 6, width = 1 / 3,
  height = 1 / 6, r = {x, 0, 0}, lengthDirection = {0, -1, 0},
  color = {255, 191, 0});
    Modelica.Mechanics.MultiBody.Visualizers.Advanced.Shape
  rail(shapeType = "box", length = 0.01, width = 4 * 1,
  height = 0.01, r = {0, 0, 0}, lengthDirection = {0, -1,
  0}, color = {195, 176, 145});
    Modelica.Blocks.Interfaces.RealInput u;
    Modelica.Blocks.Interfaces.RealOutput phi(start = pi / 6,
  fixed = true);
  equation
    (M + m) * der(der(x)) - m * 1 * der(der(phi)) * cos(phi)
  + m * 1 * der(phi) * der(phi) * sin(phi) + b * der(x) = u;
    1 * der(der(phi)) - g * sin(phi) - der(der(x))
  * cos(phi) = 0;
    xp = x - 1 * sin(phi);
    yp = 1 * cos(phi);
  end ReversPendulum;
```

We present the results of an intermediate numerical experiment. Perform a simulation without adjusting the pendulum. In this case, the pendulum flips over and we get results similar to an elliptical pendulum. In a medium with friction, we observe damped oscillations (Fig. 6.65).

The animation with the trace of the movement of the load is as shown in Fig. 6.66. We also plot the trolley displacement along the x-axis, Fig. 6.67.

Now, let us move on to controlling the pendulum.

Component: Regulator

Create the "Regulator" component, see Fig. 6.68.

Writing a program code for it is not difficult, see Example 6.1 and Fig. 6.69.

Gain can be adjusted either at the bottom of the screen, Fig. 6.70. Or when performing a numerical experiment.

And finally, combine our model.

Trolley-mounted inverted pendulum

Let us create an icon of the finished model in case we need to use a controlled pendulum in more complex systems, Fig. 6.71.

Assemble the component model from the created components, Fig. 6.72.

Run a numerical experiment with the settings of the PID controller: $K_p = -80$, $K_i = 0$, $K_d = 0$. We plot for the angular displacement of the pendulum, Fig. 6.73a, and for the linear displacement of the trolley, Fig. 6.73b. A more active damping of oscillations is observed than without a regulator, simultaneously with an

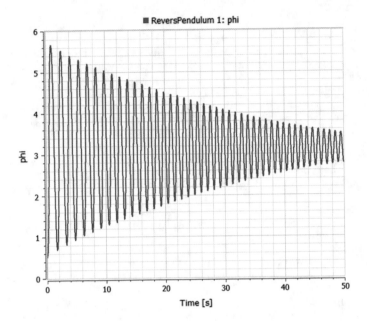

Fig. 6.65 Damped oscillations of an unregulated inverted pendulum on a trolley

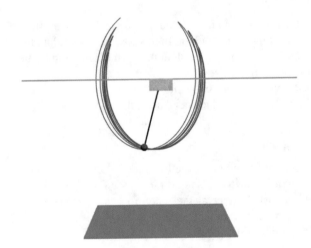

Fig. 6.66 Animation example of an unregulated inverted pendulum on a trolley

increase in the oscillation frequency, a systematic shift of the trolley in the negative direction is also noticeable.

The animation with the trace of the movement of the load is as follows, Fig. 6.74.

Change the settings of the PID controller and start the numerical experiment again. The simulation results with the settings of the PID controller: $K_p = -100$, $K_i = 0$, $K_d = -10$ are shown in Fig. 6.75. The position of the pendulum stabilizes almost

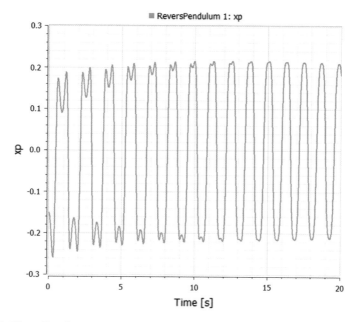

Fig. 6.67 The trolley displacement along the x-axis

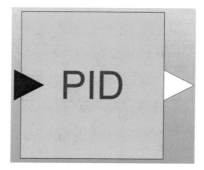

Fig. 6.68 Icon "Regulator"

instantly when the variable φ reaches values close to zero, however, the carriage is in constant motion, as can be seen from the graph x, Fig. 6.76.

When performing a numerical experiment with the settings of the PID controller: $K_p = -100$, $K_i = 4$, $K_d = -10$, stabilization of the pendulum occurs even faster without moving the equilibrium position, however, the carriage moves in the negative direction.

From the above analysis, it is clear that PID control is not quite suitable for solving the stabilization problem of this mechanical system, since the PID controller controls only the position of the pendulum relative to its equilibrium, due to the movement

```
PIDReversPendulum.PID
model PID
  parameter Real Kp = -100;
  parameter Real Kd = 0;
  parameter Real Ki = 0;
  parameter Real phi0 = 0;
  Modelica.Blocks.Interfaces.RealInput phi ¤;
  Modelica.Blocks.Interfaces.RealOutput out ¤;
  Real e, I;
equation
  e = phi - phi0;
  der(I) = e;
  out = Kp * e + Kd * der(e) + Ki * I;
  ¤;
end PID;
```

Fig. 6.69 PID controller program code

×	General	Reliability	Messages			
❐	Name	Value		Initial Value	Fixed	Description

Parameters	
Kp	-100
Kd	0
Ki	0
phi0	0

Fig. 6.70 Setting PID controller parameters

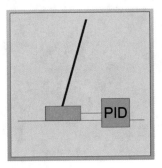

Fig. 6.71 Pictogram "controlled inverted pendulum on a cart"

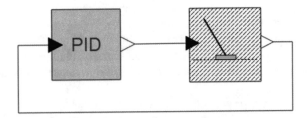

Fig. 6.72 Component hierarchical model "controlled inverted pendulum on a cart"

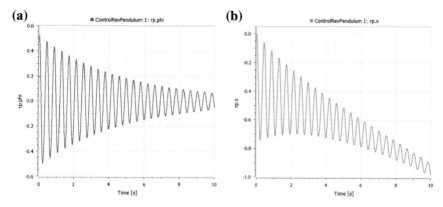

Fig. 6.73 Graphs of the angular displacement of the pendulum and the linear displacement of the trolley to the trolley with the parameters $K_p = -80$, $K_i = 0$, $K_d = 0$

Fig. 6.74 Animation example of an adjustable inverted pendulum on a trolley with $K_p = -80$, $K_i = 0$, $K_d = 0$

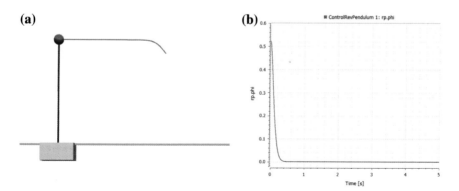

Fig. 6.75 Animation example of an adjustable inverted pendulum on a trolley and stabilization of the deflection angle with parameters $K_p = -100$, $K_i = 0$, $K_d = -10$

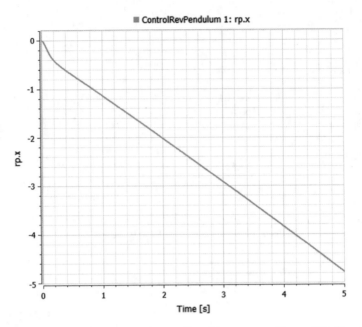

Fig. 6.76 Graph of linear displacement of the trolley with parameters $K_p = -100$, $K_i = 0$, $K_d = -10$

of the trolley. The constant movement of the trolley indicates that the system does not reach an equilibrium position, since the task of stabilizing the trolley is simply not considered. Therefore, regulation of the deviation angle alone is not enough to achieve equilibrium by the system; a regulator is needed that dynamically regulates both generalized coordinates. To do this, you can use linear-quadratic controllers (LQR) and linear-quadratic Gaussian controllers (LQG).

This task, however, is beyond the scope of this textbook and can be proposed as an independent study.

6.4   Final Remarks

Upon performing the basic tasks proposed in the textbook, we are convinced that the Wolfram SystemModeler is a fairly convenient tool for physical modeling and conducting numerical experiments. Unlike other modeling environments, Wolfram SystemModeler uses the standard Modelica physical modeling language and does not require additional components. Wolfram SystemModeler provides integration with the Mathematica package [6] for a complete process of physical, numerical simulation, and analysis of the results.

In conclusion, in order to use the capabilities of the Wolfram SystemModeler package for a wider range of tasks (beyond the scope discussed in the textbook), we present a brief comparative table with other popular modeling environments [7–9].

	SystemModeler	MapleSim*	Simulink*
Additional requirements	Mathematica is optional	*Maple required	*MATLAB required
Physical modeling			
Hierarchical modeling	Yes	Yes	Yes paid
Multi-domain modeling	Yes	Yes	Yes paid
Drag-and-drop approach, component modeling	Yes	Yes	Yes
Work with Modelica models	Yes	Possibly	No
The combination of Modelica code generation with component modeling	Yes	No	No
Creating components using symbol equations	Yes	Possibly	Possibly paid
Using C external functions	Yes	Yes	possibly
Built-in model libraries			
Electrical (analog and multi-phase)	Yes	Yes	Possibly paid
Magnetic	Yes	Yes	Yes paid
Mechanical (translational, rotational and 3D multi-component)	Yes	Yes	Possibly paid
Blocks of signals (continuous, discrete and logical signals)	Yes	Yes	Yes
State diagrams	Yes	No	Yes paid
Thermal (heat transfer and fluid flow)	Yes	Yes	Possibly paid
Numerical experiment			
Hybrid continuous discrete solution algorithms	Yes	Possibly	Yes
Finding real-time solutions	Yes	No	Yes paid

(continued)

(continued)

	SystemModeler	MapleSim*	Simulink*
Sensitivity analysis methods	Yes	Yes	No
Visualization			
Plotting any system variable with one click	Yes	No	No
Automatic 3D visualization of mechanical systems	Yes	Yes	No
Configurable visualization environment			
2D and 3D graphics language	Yes	Possibly	Yes
Advanced 3D graphics (lighting, transparency, etc.)	Yes	Possibly	Yes
Two-dimensional and three-dimensional animation	Yes	Possibly	Possibly
Standard formats (.avi and .mov)	Yes	No	Possibly
Direct interactivity	Yes	No	No
Analysis and design			
Numerical playback control	Yes	Yes	Yes
Analysis of model equations	Yes	Yes	No
Checking model balance	Yes	Yes	Yes paid
Management systems design	Yes	Yes paid	Yes paid
Model calibration	Yes	Yes	Yes
System optimization	Yes	Yes	Yes
Dynamic charts	Yes	No	No
Reliability analysis	Yes	No	No

To simulate complex systems, many different visual modeling environments have been created. Using these environments allows you to quickly build a model, automatically get executable code, and using the built-in analysis tools go directly to the study of model properties.

It is obvious that the simultaneous use of the language of mathematics and any visual modeling environment in computer modeling of complex systems is an effective method for solving a wide range of engineering problems.

The answer to the question related to the choice of a specific environment and modeling language depends on the type of application and user preferences.

The tutorial examined the Wolfram SystemModeler modeling environment, based on Modelica. According to the authors, this environment, having wide functional capabilities and visualization tools, is a universal tool for studying applied problems.

References

1. Yu.B. Kolesov, Yu.B. Senichenkov, Matematicheskoe modelirovanie gibridnyh dinamicheskih sistem. – Izdatelstvo Politekhnicheskogo universiteta, SPb (2014)
2. Yu.B. Kolesov, Yu.B. Senichenkov, Modelirovanie sistem. Ob'ektno-orientirovannyj podhod. – BHV-Peterburg, SPb (2012)
3. Yu.B. Kolesov, Yu.B. Senichenkov, Matematicheskoe modelirovanie slozhnyh dinamicheskih sistem. – Izdatelstvo SPbPU, SPb (2018)
4. A. Urquía, C. Martín-Villalba, M.A. Rubio, V. Sanz, Simulation practice with Modelica. – Editorial UNED, Madrid (2018)
5. D. Liberzon, Switching in Systems and Control. – Birkhäuser, Basel (2003)
6. Wolfram Mathematica: https://www.wolfram.com/mathematica/
7. Matlab Simuink: https://www.mathworks.com/
8. Maple: https://www.maplesoft.com/
9. Wolfram SystemModeler: https://www.wolfram.com/system-modeler/